David Vetsch

Numerical Simulation of Sediment Transport with Meshfree Methods

David Vetsch

Numerical Simulation of Sediment Transport with Meshfree Methods

Basics, Methods and Applications

Südwestdeutscher Verlag für Hochschulschriften

Impressum / Imprint
Bibliografische Information der Deutschen Nationalbibliothek: Die Deutsche Nationalbibliothek verzeichnet diese Publikation in der Deutschen Nationalbibliografie; detaillierte bibliografische Daten sind im Internet über http://dnb.d-nb.de abrufbar.
Alle in diesem Buch genannten Marken und Produktnamen unterliegen warenzeichen-, marken- oder patentrechtlichem Schutz bzw. sind Warenzeichen oder eingetragene Warenzeichen der jeweiligen Inhaber. Die Wiedergabe von Marken, Produktnamen, Gebrauchsnamen, Handelsnamen, Warenbezeichnungen u.s.w. in diesem Werk berechtigt auch ohne besondere Kennzeichnung nicht zu der Annahme, dass solche Namen im Sinne der Warenzeichen- und Markenschutzgesetzgebung als frei zu betrachten wären und daher von jedermann benutzt werden dürften.

Bibliographic information published by the Deutsche Nationalbibliothek: The Deutsche Nationalbibliothek lists this publication in the Deutsche Nationalbibliografie; detailed bibliographic data are available in the Internet at http://dnb.d-nb.de.
Any brand names and product names mentioned in this book are subject to trademark, brand or patent protection and are trademarks or registered trademarks of their respective holders. The use of brand names, product names, common names, trade names, product descriptions etc. even without a particular marking in this works is in no way to be construed to mean that such names may be regarded as unrestricted in respect of trademark and brand protection legislation and could thus be used by anyone.

Coverbild / Cover image: www.ingimage.com

Verlag / Publisher:
Südwestdeutscher Verlag für Hochschulschriften
ist ein Imprint der / is a trademark of
AV Akademikerverlag GmbH & Co. KG
Heinrich-Böcking-Str. 6-8, 66121 Saarbrücken, Deutschland / Germany
Email: info@svh-verlag.de

Herstellung: siehe letzte Seite /
Printed at: see last page
ISBN: 978-3-8381-3294-5

Zugl. / Approved by: Zurich, ETH Zurich, Diss., 2011

Copyright © 2012 AV Akademikerverlag GmbH & Co. KG
Alle Rechte vorbehalten. / All rights reserved. Saarbrücken 2012

Acknowledgements

The author of this doctoral thesis likes to thank his supervisor Prof. Dr. R. Boes (ETH Zurich) and the co-examiners Prof. Dr. P. Rutschmann (TU Munich) and Dr. R. Fäh (ETH Zurich). Furthermore, the author is very grateful to the co-examiner Dr. F. Fleissner (TU Stuttgart), who provided his software package *Pasimodo* for academic use and to Mrs. A. Lehnart for her substantial contributions to the software. In addition, the author would like to express his gratitude to Prof. Dr. K. Hutter for the critical review and his valuable suggestions and Dr. U. Keller for his editorial contributions. The present research work was financially supported by the Swiss Federal Office for the Environment and its support has been much appreciated.

Contents

Abstract ix

Kurzfassung xi

1 INTRODUCTION 1
   1.1 Motivation 1
   1.2 Contribution of this Thesis 2

2 LITERATURE REVIEW 3
   2.1 Modelling of Bed Load Transport 3
      2.1.1 Definition of Bed Load Transport 3
      2.1.2 Incipient Motion 4
      2.1.3 Empirical Relations for Bed Load Transport 13
      2.1.4 Numerical Modelling of Sediment Transport in Rivers 16
   2.2 Numerical Methods and Investigations 19
      2.2.1 Numerical Methods for Fluid Flow 19
      2.2.2 Modelling of Granular Material 23
      2.2.3 Simulation of Fluid and Sediment Particles 24

3 PHYSICS OF FLUIDS AND RIGID BODIES 27
   3.1 Basic Description of a Conservation Law 27
   3.2 Governing Equations for Fluid Flow 28
      3.2.1 Conservation of Mass 28
      3.2.2 Conservation of Momentum 29
      3.2.3 Conservation of Energy 30
      3.2.4 Simplifications 31
   3.3 Motion of Rigid Bodies 34
      3.3.1 Equations of Motion 34
      3.3.2 Applied Forces and Torques 35
      3.3.3 State of a Rigid Body 36
   3.4 Fluid and Rigid Bodies 37
      3.4.1 Hydrostatic Pressure and Buoyancy 38
      3.4.2 Hydrodynamic Forces 40

4 NUMERICAL METHODS 45
   4.1 Introduction 45
      4.1.1 Methodology 45
      4.1.2 Software Framework *Pasimodo* 46
   4.2 Smoothed Particle Hydrodynamics 48

	4.2.1 Introduction	48
	4.2.2 Representation of Fluids by Particles	49
	4.2.3 Governing Equations	52
	4.2.4 Enhancements	54
	4.2.5 Initial Conditions	59
	4.2.6 Boundary Conditions	59
	4.2.7 Time Integration and Solution Algorithm	61
	4.2.8 Considerations of Accuracy	63
4.3	Discrete Element Method	64
	4.3.1 Basic Concepts	64
	4.3.2 Penalty Force Models	66
	4.3.3 Modelling of Collisions	74
	4.3.4 Friction	77
	4.3.5 Time Integration	82
	4.3.6 Considerations of Accuracy	84
	4.3.7 Choice of Appropriate Simulation Models and Parameters	88
4.4	Fluid-Structure Interaction	91
	4.4.1 Normal Forces	91
	4.4.2 Friction	96
	4.4.3 Damping	99
	4.4.4 Time Integration and Solution Algorithm	100
5	**MODEL CALIBRATION AND VALIDATION**	**101**
5.1	Introduction	101
5.2	Buoyancy	102
	5.2.1 Configurations	102
	5.2.2 Boundary and Initial Conditions	105
	5.2.3 Results	107
5.3	Settling Velocity	112
	5.3.1 Configurations	113
	5.3.2 Boundary and Initial Conditions	115
	5.3.3 Results	115
5.4	Open Channel Flow	124
	5.4.1 Configurations	124
	5.4.2 Boundary and Initial Conditions	125
	5.4.3 Results	126
5.5	Discussion	129
6	**APPLICATION TO SEDIMENT TRANSPORT**	**131**
6.1	Introduction	131
6.2	Scour Caused by a Freefalling Water Jet	131
6.3	Clear-Water Scour at Bridge Pier	135
7	**CONCLUSION**	**139**
7.1	Summary	139
7.2	Recommendations for Future Research	141

Bibliography		143
List of Symbols		159
List of Acronyms		167
Appendix		169
A.1	Material Properties	169
A.1.1	Stiffness and Young's Modulus	169
A.1.2	Poisson Ratio	169
A.1.3	Viscosity	170
A.1.4	Friction Coefficients of Granular Materials	170
A.1.5	Summary	172
A.2	General Definitions for DEM	173
A.3	MLJ Potential for Sphere to Sphere Interaction	175
A.4	Different Forms of Lennard-Jones Potentials	177
A.4.1	Form Depending on Equilibrium Distance	177
A.4.2	Form According to Monaghan	177
A.4.3	Form According to Müller	177
A.4.4	Approximations	179
A.5	Vector Projection	179

Abstract

The assessment of sediment transport and involved processes is a major issue in hydraulic and river engineering. The estimation of erosion and deposition in terms of transport rates is crucial for the dimensioning of river works, such as river corrections or local river widenings, and for an appropriate protection of hydraulic structures. In recent years, also ecological aspects in combination with morphological changes of the river bed have become more and more important. The common approaches for the determination of sediment transport rates are mostly based on empirical relations, which were obtained by evaluation of field measurements or laboratory experiments. Because of their empirical nature, these rather simple approaches are useful for the practical estimation of transport rates based on averaged quantities. However, their application is generally subject to more or less laborious and expensive calibration. Furthermore, they are not suitable to study the generally complex sediment transport processes since the empirical approaches are not able to describe the underlying physics in detail. Hence, numerous experimental investigations by various research groups using state-of-the-art measuring techniques have been carried out in the past to gain an in-depth insight into the processes involved in sediment transport. To complement the corresponding findings also numerical investigation into the subject would be desirable. However, numerical models which are able to reproduce and to resolve the involved processes are not very common, since they would have to imply the rather complex fluid-sediment interaction.

In the present work, a numerical model which is based on two meshfree particle methods is presented. The fluid is modelled by a continuum approach which is discretised by the <u>S</u>moothed <u>P</u>article <u>H</u>ydrodynamics (SPH) method. The sediment particles are represented by the <u>D</u>iscrete <u>E</u>lement <u>M</u>ethod (DEM), where the interactions between the discrete sediment grains are modelled by a force law, which is also able to account for various kinds of friction. A similar approach is applied to the interaction between the fluid and sediment particles. The definition of the interface and the exchange of forces between the fluid and sediment grains are inherent to the applied approaches. Thus, the application of special techniques to describe a movable or deformable interface as used for grid-based methods is not necessary.

For the application of the combined methods, two different modelling approaches are pursued. On the one hand, the fluid particles are chosen distinctly smaller than the sediment particles to simulate detailed interaction forces. The study of the influence of the particle resolution, a hydrostatic and a dynamic experiment, namely the simulation of buoyancy effects and the determination of the settling velocity, have been carried out. The simulations show convergence of the results for increasing particle resolution and turned out to be a reliable concept to validate the chosen numerical approaches. On the other hand, the fluid particles are chosen of similar size or larger than

Abstract

the sediment particles. Due to the less detailed resolution of the fluid forces acting on a solid particle, the model parameters have to be calibrated to match the desired sediment transport processes; this applies to the spatial as well as to the temporal scale. Furthermore, the solid particles may no longer only represent a single sediment grain, but rather a small volume of sediment or a chunk of soil. This modelling approach was successfully applied to scour caused by a freefalling water jet and to clear-water scour at a bridge pier.

The satisfying simulation results demonstrate the potential of the presented model for the detailed investigation of sediment transport processes as well as for complex practical applications. However, besides some shortcomings which still are to overcome, the main restriction of the applied method is their computational cost which makes the use of high performance computing infrastructure inevitable. Nevertheless, the combination of the presented numerical methods is a promising modelling approach, which may serve as an appropriate simulation tool for many hydraulic and river engineering problems in the future.

Kurzfassung

Die Erfassung von Sedimenttransport und den dabei involvierten Prozessen ist eine wichtige Aufgabe im Wasser- und Flussbau. Die Bestimmung von Erosion und Ablagerungen im Sinne von Transportmengen ist entscheidend für die Dimensionierung von flussbaulichen Massnahmen, wie etwa Flusskorrekturen oder lokale Flussaufweitungen, sowie für eine angemessene Auslegung von Massnahmen zum Schutz wasserbaulicher Strukturen. Ebenfalls hat in den letzten Jahren die Bedeutung ökologischer Aspekte in Kombination mit morphologischen Veränderungen des Gerinnes zugenommen. Die gängigen Ansätze zur Bestimmung der Sedimentfracht beruhen vorwiegend auf empirischen Formeln, welche anhand der Auswertung von Feldmessungen oder Laborexperimenten hergeleitet wurden. Aufgrund ihrer empirischen Natur sind diese eher einfachen Ansätze gut geeignet für eine praktische Abschätzung des Sedimenttransports basierend auf gemittelten Grössen. Jedoch ist mit deren Anwendung grundsätzlich eine mehr oder weniger aufwändige Kalibrierung verbunden. Des Weiteren sind diese Ansätze zu einer Untersuchung der unterschiedlichen und oftmals komplizierten Vorgänge bei Sedimenttransport nicht geeignet, da die empirischen Ansätze die grundlegenden physikalischen Vorgänge nicht wiedergeben. Daher wurde in der Vergangenheit eine Vielzahl an experimentellen Untersuchungen von verschiedenen Forschungsgruppen unter Verwendung moderner Messtechniken durchgeführt, um einen vertieften Einblick in die dem Sedimenttransport zugrundeliegenden Prozesse zu erhalten. Zur Ergänzung der daraus entstandenen Erkenntnisse wären ebenfalls numerische Untersuchungen wünschenswert. Jedoch sind entsprechende numerische Modelle, welche fähig sind, die wesentlichen Vorgänge aufzulösen und wiederzugeben, eher selten, da ein solches Simulationsmodel auch Ansätze für die komplexe Fluid-Struktur-Koppelung beinhalten muss.

In der vorliegenden Arbeit wird ein numerisches Modell basierend auf zwei gitterfreien Partikelmethoden präsentiert. Dabei wird das Fluid als Kontinuum mittels der *Smoothed-Particle-Hydrodynamics* (SPH) Methode diskretisiert. Das Sediment wird mit der *Discrete-Element-Methode* (DEM) wiedergegeben, wobei die Interaktion zwischen den einzelnen Sedimentkörnern mittels eines Kraftgesetztes modelliert wird, welches ebenfalls verschiedene Arten von Reibung mit einschliesst. Ein gleichartiger Ansatz wird für die Interaktion zwischen dem Fluid und den Sedimentpartikeln verwendet. Die Bestimmung der entsprechenden Kontaktflächen und der Austausch von Kräften zwischen dem Fluid und den Sedimentkörnern ist dem gewählten Ansatz inhärent. Daher ist die Verwendung von speziellen Verfahren zur Behandlung einer beweglichen oder verformbaren Kontaktfläche, wie dies bei gitterbasierten Verfahren der Fall ist, nicht notwendig.

Zur Anwendung der kombinierten Methoden wurden zwei unterschiedliche Modellierungsansätze verfolgt. Einerseits wurden die Fluidpartikel deutlich kleiner als die Sedimentpartikel gewählt, um

Kurzfassung

die detaillierten Interaktionskräfte zu simulieren. Um den Einfluss der Partikelauflösung zu analysieren, wurden ein hydrostatisches und ein hydrodynamisches Experiment, d.h. die Simulation des Auftriebseffekts und die Bestimmung der Sinkgeschwindigkeit, durchgeführt. Die Simulationen zeigen die Konvergenz der verwendeten Verfahren mit zunehmender Partikelauflösung. Zudem stellte sich heraus, dass die gewählte Vorgehensweise ein zuverlässiges Konzept zur Validierung solcher Verfahren ist. Andererseits wurden Versuche durchgeführt, bei welchen die Fluidpartikel von der gleichen Grösse oder grösser als die Sedimentpartikel sind. Aufgrund der dabei weniger detaillierten Auflösung der Interaktionskräfte müssen die zugehörigen Modellparameter kalibriert werden, um die gewünschten Vorgänge des Sedimenttransports wiederzugeben. Die Kalibrierung betrifft sowohl die räumliche als auch die zeitliche Skala der Prozesse. Des Weiteren entsprechen in diesem Fall die Sedimentpartikel nicht mehr einem einzelnen Korn, sondern eher einem kleinen Sedimentvolumen oder einer Handvoll Substrat. Dieser Modellansatz wurde erfolgreich auf die Kolkbildung durch einen auftreffenden Freistrahl und die Kolkvorgänge bei einem Brückenpfeiler angewendet.

Die zufriedenstellenden Simulationsresultate demonstrieren das Potential des verwendeten Modells zur detaillierten Untersuchung von Sedimenttransportvorgängen und für komplexe praktische Anwendungen. Nebst einigen überwindbaren Schwächen des Modells stellt jedoch der erforderliche Rechenaufwand das grösste Hindernis dar, welches nur durch Einsatz von Hochleistungsrechnern überwunden werden kann. Gleichwohl ist die Kombination der präsentierten numerischen Verfahren ein vielversprechender Modellierungsansatz, welcher zukünftig als zweckdienliches Simulationswerkzeug für verschiedene Fragestellungen im Bereich des Fluss- und Wasserbaus dienen kann.

1 INTRODUCTION

1.1 Motivation

Sediment transport in rivers is studied for more than a century with the goal to estimate transport rates and the corresponding change of the river-bed topography. Unforeseen aggradations of the river bed may compromise flood safety due to the reduction of the flow section. In contrast, degradations and local erosion can lead to destabilisations of embankments or bridge piers, for example. Thus, a reliable prediction of the erosion and deposition of sediment in an alluvial river is an important issue related to river correction works or the design of hydraulic structures. In recent years, ecological aspects in combination with revitalisation measures like local river widenings reveal new challenges for river engineering, especially with regard to sediment transport.

Investigations of river morphology have mainly an experimental background. The processes involved in sediment transport, as the inception of motion, the transport itself and the deposition of sediment, are usually reduced to empirical relations and are combined in the form of a transport formula. Especially the common concept of incipient motion, where motion of sediment depends on a threshold condition, has to be questioned. Alternative approaches based on probability distributions used to describe the state of the sediment seem to be more reliable, since their concept corresponds to the natural continuous motion of sediment. Furthermore, the driving forces acting on the sediment, which actually cause the transport, are usually derived from averaged flow quantities. These approaches are useful and of great importance for engineering practice, but they only allow for the determination of a temporally and spatially averaged sediment transport.

For river engineering problems, where the morphological development plays an important role, a variety of numerical tools exists. These are able to simulate sediment transport from a local to a regional scale with satisfying accuracy as far as sufficient data for their calibration is available. By the application of modern numerical tools it is nowadays possible to resolve the flow field, i.e. the water phase, in detail. However, depending on the resolved scales, the gained advantage will be lost due to the rather approximate approach for sediment transport, i.e. the solid phase of the water-sediment mixture flow.

In the last decades, many researchers tried to overcome the shortcomings of physics in the common approaches, however, with limited success. Despite investigations using state-of-the-art measuring techniques and providing an in-depth view of acting forces at the sediment bed, a reasonable approach, which does not need calibration but is still convenient for practical application, does not seem to be available in the near future. However, such kinds of investigations highlight the complexity of the involved processes and the sediment transport per se. Furthermore, the detailed experimental data may serve for the validation of advanced numerical models.

1 Introduction

Because of the availability of increasing computational resources, the application of numerical models for the investigation of the mechanics of sediment transport becomes more and more popular. Such numerical tools are rather sophisticated, since they have to be able to model the interaction between the fluid and the sediment grains as well as the interactions between the grains themselves. Furthermore, such models have to include friction to correctly reproduce the constitutional behaviour of the sediment and the different modes of bed load transport, as sliding, rolling and saltating. One of the main challenges in developing such approaches is the appropriate modelling of the movable interfaces between the fluid and the sediment grains and the exchange of forces. Although several different numerical techniques exist which are suitable for such problems, they often have deficits concerning efficiency or accuracy. Furthermore, many common numerical approaches for the simulation of fluid flow use computational grids for the spatial discretisation, which may reduce the flexibility for the modelling of arbitrary geometries and lead to quite complex schemes for movable boundaries. However, when it comes to three dimensional applications, the main handicap of these approaches is the computational expense necessary to obtain qualitatively good results, and the use of high performance computing seems to be inevitable.

1.2 Contribution of this Thesis

The main goal of this thesis is to evaluate a novel modelling approach and the underlying numerical methods which are able to simulate sediment transport and reproduce the involved processes in detail. To reduce the complexity of this challenging task, the primary focus of this work is on bed load transport. Since the involved physical processes rely on fluid and rigid body dynamics, numerical discretisation techniques are applied, which account for the distinct characteristics of these disciplines and which allow for flexible modelling of fluid-structure interaction. Thus, the application of two meshfree particle methods is considered. These are able to model the different properties of the fluid and the sediment as well as their interaction without the need for a computational grid. Furthermore, this hybrid approach allows for the description of the processes of bed load and the corresponding transport modes by discrete forces. The successful application of the model to various problems shows the potential of this approach for the numerical simulation of bed load transport. The model is a suitable numerical research tool and may serve for future investigations, especially with regard to increasing computing power.

2 LITERATURE REVIEW

2.1 Modelling of Bed Load Transport

2.1.1 Definition of Bed Load Transport

Sediment transport in rivers can be classified by two types referring to their transport mechanism, *suspended sediment* and *bed load transport*. Suspended sediment transport, as the term denotes, is the part of the load which is carried in suspension by the movement of the water. Thus, the transported sediment mainly consists of fine material like sand or silt.

The mechanism of bed load transport is described by processes occurring in the upper-most layer of the river bed. Sediment grains are moved in different forms due to stream forces or strikes of other grains in motion (Bagnold (1941)). The transport modes are comparable with Aeolian transport which can be observed at sand dunes in deserts; grains move in flow direction by saltating, or, which is less usual, by rolling or even by sliding along the bed (Fig. 2-1). The distinction between transport in the form of saltation or in suspension is not obvious. Bagnold (1973) defines transport of a solid in suspension as a state in which the excess weight of the solid is compensated by a random succession of upward impulses due to eddy currents of fluid turbulence moving upwards relative to the bed. Therefore, the solid may remain out of contact with the bed for an indefinite period depending on the random nature of turbulence. In contrast, saltation as well as bed load transport in general may be characterised as motion with successive contacts between the solid and the bed.

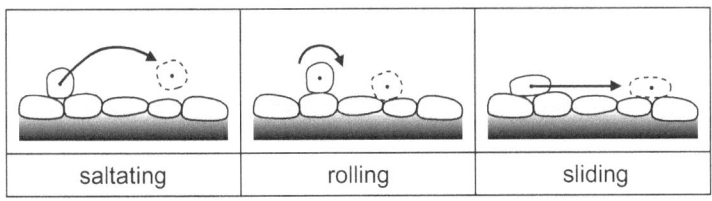

Fig. 2-1: Modes of bed load transport

It should be noted that the early approach of du Boys (1879), which assumes that bed load occurs as stacked layers moving in a "carpet like" form has nothing in common with today's view (see e.g. Yalin (1977)). The approach of du Boys is rather applicable to granular flow, which can be observed in fluidized beds or high-concentration granular-liquid mixtures (Armanini *et al.* (2005)) under shear flow.

2 Literature Review

The grain size of transported material in alpine rivers covers a wide range and is described by the smaller diameter d_s of a non-uniformly shaped grain. The size varies from sand (0.062 < d_s < 2 mm) over gravel (2 < d_s < 60 mm) to stones and small rocks. Finer sediment than the sand fraction is termed wash load, which is considered to be moved in suspension only.

Church (2006) gives an overview on the different sediment transport regimes, on the categorization of fluvial sediments as well as on the relation of bed load transport and morphology in alluvial rivers. A quantitative distinction between bed load and suspended load can be found in Murphy and Aguirre (1985).

The focus of this thesis is on the process of bed load transport in alluvial gravel-bed rivers, typical examples of which are the Swiss rivers Aare, Thur or the Alpine Rhine and many other mountainous rivers worldwide.

2.1.2 Incipient Motion

The topic of incipient motion - the onset of transport of sediment - has been studied by many researchers in the last hundred years or so. In most cases, the goal was to define a threshold for sediment motion which is an essential premise for the estimation of sediment transport in alluvial rivers. The main motivation for the investigations was and still is the development of a transport relation to asses bed load discharge in rivers that serves as essential tool for river engineering works. The methodology to find a criterion for the threshold is usually based on theoretical investigations or visual observations as well as measured reference bed load transport rates, acquired in a laboratory flume or in a natural river.

Some results of investigations into threshold of sediment motion are shown in Fig. 2-2, whereas the abscissa denotes the grain Reynolds number of the sediment and the ordinate indicates the threshold stream force in terms of critical shear stress. Based on the consideration of equilibrium of moments of a spherical grain and dimensional analysis, Shields (1936) proposed to express the dimensionless critical shear stress τ_c^* as a function of the grain or particle Reynolds number. This finding may be formulated in terms of critical values (see e.g. Yalin (1977)) for a grain of size d_s as

$$\tau_c^* = \frac{\tau_c}{\triangle \rho g d_s} = f(\text{Re}_c^*), \quad \text{Re}_c^* = \frac{u_{*c} d_s}{\nu}, \tag{2.1}$$

where $\triangle \rho = (\rho_s - \rho_f)$ is the density difference between sediment (subscript s) and fluid (subscript f), g is the gravitational acceleration, Re_c^* is the critical particle Reynolds number, ν is the kinematic viscosity. The dimensionless critical shear stress τ_c^* (also denoted as the critical Shields parameter θ_c) and the critical shear velocity u_{*c} were determined by observations in a laboratory flume.

However, the definition of the point of inception is not clear and varies considerably among the various studies. This means that in practical cases of turbulent flow there is no single criterion for the beginning of movement of sediment. Buffington and Montgomery (1997) give an extensive review on the issue. There is also a large data collection available by Brownlie (1985) and Brownlie and Brooks (1981).

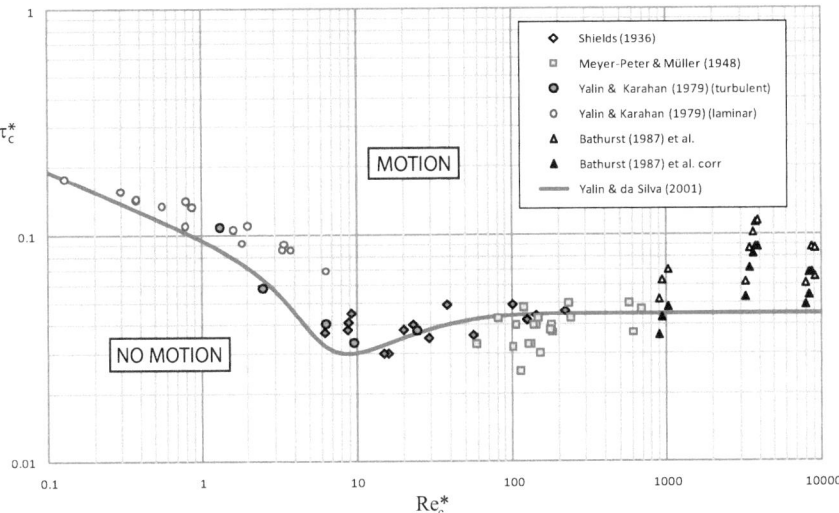

Fig. 2-2: Selected results of investigations into incipient motion

Some principal aspects of the concept of incipient motion or critical shear stress derived from empirical investigations are given by Paintal (1971):

- a distinct condition for the beginning of movement does not exist, i.e. there is no single value of bed shear stress below which not a single particle will move;
- bed load transport in the proximity of the so called critical shear stress is governed by certain laws;
- a limiting shear stress for a bed material can be defined below which the bed load transport rate is of no practical importance.

Consequently, in engineering practice the rate of sediment transport is calculated with empirically based transport equations (see chapter 2.1.3) which are usually defined as a function of a certain threshold. From a physical point of view it is obvious to express a threshold condition in terms of stream force. Thus, approaches based only on the mean flow velocity seem not to be reasonable because they do not account for flow depth and turbulence. Thus, the criterion for incipient motion is usually determined by threshold quantities like the critical bed shear stress τ_c, the critical shear velocity u_{*c} or the amount of the critical lift force $F_{L_c} = \left| \vec{F}_{L_c} \right|$. Dey and Papanicolaou (2008) provide a review on the different concepts. The threshold conditions with respect to the effective bed shear stress τ_b, velocity u_* or amount of the lift force $F_L = \left| \vec{F}_L \right|$ are shown in Tab. 2-1.

Tab. 2-1: Different concepts for threshold of bed load transport

Threshold Condition	Shear Stress	Shear Velocity	Lift Force
Bed Load Transport	$\tau_b \geq \tau_c$	$u_* \geq u_{*c}$	$F_L \geq F_{L_c}$
NO Bed Load Transport	$\tau_b < \tau_c$	$u_* < u_{*c}$	$F_L < F_{L_c}$

2.1.2.1 Conventional Threshold Criterion

One of the most cited works on sediment threshold was carried out by Shields (1936). Within the scope of his doctoral thesis he developed the Shields diagram, which is still widely used in engineering practice. His results are based on theoretical considerations and laboratory flume experiments with steady flow conditions and near-uniform noncohesive grains. Unlike previous researchers, he correlated the particle (or grain) Reynolds number Re^* to the dimensionless shear stress τ_b^* by application of dimensional analysis. Shields obtained his data mostly by extrapolating curves of sediment transport rates versus applied shear stress to the zero transport condition. The results of Shields experiments are approximated by a single curve shown in the Shields diagram Fig. 2-3 and can be expressed in simplified terms in the following way:

$$\begin{aligned} &\text{Re}_c^* \leq 2: \ \tau_c^* \approx 0.1/\text{Re}_c^*, \\ &\text{Re}_c^* = 10: \ \tau_c^* \approx 0.03, \\ &\text{Re}_c^* \geq 500: \ \tau_c^* \approx 0.056. \end{aligned} \qquad (2.2)$$

Since the original work of Shields (1936), many researchers acquired additional data to cover a wider range of Re_c^* and to arrive at an improved threshold condition (see Fig. 2-2 for selected results). For non-cohesive sediments mainly consisting of gravel, Meyer-Peter and Müller (1948) ran laboratory experiments at VAW[1] to develop a formula for bed load transport in gravel bed rivers like the Alpine Rhine, and Bathurst et al. (1987) studied the threshold condition for steep mountain streams, for example. To point out the sensitivity of the approach for the calculation of the dimensionless shear stress, the results of the latter are drawn in Fig. 2-2 based on their formula and according to that of Shields (denoted as "Bathurst (1987) corr"). A similar but more refined approach including the ratio of flow depth h_f to the roughness height k_s, h_f/k_s, is presented by Bettess (1984). Furthermore, a new theoretical model proposed by Recking (2009) for small relative flow depth h_f/d_s and steep slopes is able to adequately reproduce the increase of the critical Shields stress with increasing slope.

---

[1] Laboratory of Hydraulics, Hydrology and Glaciology (VAW) of the Swiss Federal Institute of Technology (ETH) in Zurich

2.1 Modelling of Bed Load Transport

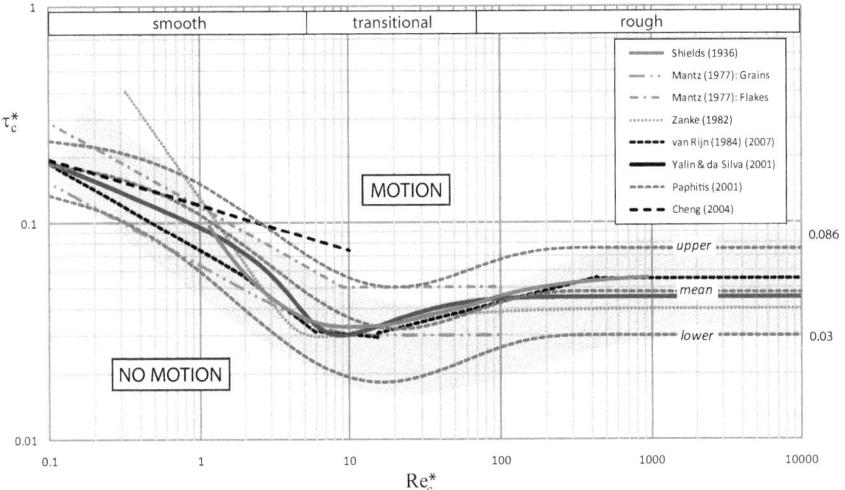

Fig. 2-3: Shields diagram with threshold curves of experimental data acquired by different researchers. The shaded area covers the data collected by Buffington and Montgomery (1997). The empirical threshold curves presented by Paphitis (2001) are denoted by "upper", "mean" and "lower" (see section 2.1.3).

In agreement with the results of Shields, Meyer-Peter and Müller (1948) found that the lower limit for absolute rest is about $\tau_c^* = 0.03$. Furthermore, they determined the threshold value for gravel with grain diameters ranging from 5 to 30 mm by extrapolation of their data to a zero transport state and obtained $\tau_c^* = 0.047$.

Investigations into fine sediments such as sand and silt, with non-cohesive or cohesive response, cover the left part of the Shields diagram. Mantz (1977) studied the transport of fine cohesionless grains and flakes in a laboratory flume based on a flat bed at condition of maximum stability arriving at the "extended Shields diagram". Mantz defined a flat bed of maximum stability as one for which a small stress increment above that for incipient transport will cause a change in bed configuration, i.e. the development of bed forms. This feature is characteristic for the left part of the Shields diagram. Shields also observed such bed forms and annotated them directly above the experimental points of his diagram. For sand bed rivers the kind of bed form varies from ripples to bars, dunes and anti dunes depending on the flow conditions (van Rijn (1984)).

As noted above, the Shields diagram depends on the dimensionless critical particle Reynolds number Re_c^* and the critical shear stress τ_c^*. Both quantities depend on the the shear velocity u_{*c} that can either be derived from the energy slope or the mean flow velocity of a cross section or can be approximated based on measurements with modern apparatus. The bottom shear stress is related to the shear velocity as

$$\tau_b = \rho_f g h_f S_e = \rho_f u_*^2 \,, \tag{2.3}$$

7

2 Literature Review

where S_e is the slope of the energy grade line which for uniform flow is equal to the bed slope S_b. For flume conditions where the friction of the side walls can be neglected due to very small roughness, h_f is the flow depth (also valid for very wide channels). Based on the shear stress velocity u_*, the steady vertical velocity profile for channel flow can be determined using the general law for wall-bounded turbulent flows (see e.g. Schlichting *et al.* (2000))

$$\frac{\bar{u}(z)}{u_*} = \frac{1}{\kappa}\ln\frac{z}{k_s} + \frac{1}{\kappa}\ln\mathrm{Re}^* + C_1(\mathrm{Re}^*) = \frac{1}{\kappa}\ln\frac{z}{k_s} + C_2(\mathrm{Re}^*) , \qquad (2.4)$$

where $\kappa \approx 0.4$ is the von Kármán constant, z is the distance from the wall and $\bar{u}(z)$ is the averaged flow velocity parallel to the wall at z (an example is depicted in Fig. 3-5 on page 42). The constants C_1 and C_2 (see Tab. 2-2) have to be determined experimentally and can be found in Nikuradse (1933). By integration of equation (2.4) over the flow depth h_f one may deduce the average flow velocity $\bar{u}_m = \bar{u}(z_m)$ and the location of its centre: $z_m = e^{-1}h_f \approx 0.368 h_f$. A detailed derivation is provided by Yalin (1977) for example. Furthermore, the wall roughness is expressed in terms of Nikuradse's sand roughness k_s, which for a flat bed covered by uniform spheres is equal to the grain diameter d_s. Accordingly, the grain or roughness Reynolds number can be written as

$$\mathrm{Re}^* = \frac{u_* d_s}{\nu} = \frac{u_* k_s}{\nu} . \qquad (2.5)$$

Schlichting (1936) provides roughness values for several kinds of surfaces, termed "equivalent sand roughness" $k_{s,eq}$, that are more global and suitable for practical use.
According to the grain Reynolds number, or the sand roughness height, the roughness flow regimes for spherical particles can be divided into three sections as summarised in Tab. 2-2 by Schlichting *et al.* (2000), which also leads to a meaningful subdivision of the Shields diagram, as depicted in Fig. 2-3.

Tab. 2-2: Roughness flow regimes for spherical particles

$0 \leq \mathrm{Re}^* \leq 5$	$C_1 \approx 5$	smooth	Roughness elements are completely covered by the laminar boundary layer, i.e. the viscous sublayer
$5 < \mathrm{Re}^* < 70$	$C(k_s)$	transitional	
$70 \leq \mathrm{Re}^*$	$C_2 \approx 8$	rough	Roughness elements are completely exposed to turbulent flow and the laminar boundary layer almost vanishes. From original work of Nikuradse $C_2 = 8.48$

Besides the investigations and considerations for turbulent channel flow presented above, Yalin and Karahan (1979) extended the Shields diagram based on measurements with focus on viscous dominated flow conditions. They carried out flume experiments with a glycerine-water mixture to obtain laminar flow and with water for turbulent flow. For laminar flow with a free surface the law of the wall reads

$$\frac{\overline{u}(z)}{u_*} = \text{Re}^* \left[\frac{z}{h_f} \left(1 - 0.5 \frac{z}{h_f} \right) \right], \qquad (2.6)$$

and the value of the average flow velocity $\overline{u}_m$ can be evaluated at $z_m = (1 - \sqrt{3}/3)h_f \approx 0.422 h_f$. Their results (depicted in Fig. 2-2) led to the hypothesis, that in laminar flows there is a distinct curve for the inception of sediment transport. More recent contributions to bed load transport in laminar flows by Pilotti and Menduni (2001) confirm this hypothesis. Their results are depicted in Fig. 2-3 in the form of a linear fit by Cheng (2004).

2.1.2.2 Explicit Formulas for Threshold Curves

For engineering purposes, one drawback of the Shields diagram is the inter-dependency of the diagram axes, i.e. the shear stress velocity u_* appears on both axes. This implies that for given stream forces and grain material the critical shear stress has to be determined by iteration, which can be laborious. To overcome this fact, some researchers provide approximations of the experimental data by explicit formulas, presented below. The resulting single curves are depicted in Fig. 2-3).

For numerical models, the following expressions by van Rijn (1984) and van Rijn (2007), are frequently used to calculate the critical shear stress

$$\begin{aligned}
\tau_c^* &= 0.115(D^*)^{-0.5} & \text{for } D^* < 4 \,, \\
\tau_c^* &= 0.14(D^*)^{-0.64} & \text{for } 4 \leq D^* < 10 \,, \\
\tau_c^* &= 0.04(D^*)^{-0.1} & \text{for } 10 \leq D^* < 20 \,, \\
\tau_c^* &= 0.013(D^*)^{0.29} & \text{for } 20 \leq D^* \leq 150 \,, \\
\tau_c^* &= 0.055 & \text{for } D^* > 150 \,,
\end{aligned} \qquad (2.7)$$

where D^* is the dimensionless grain diameter

$$D^* = \Xi = \left(\frac{(\text{Re}^*)^2}{\tau_c^*} \right)^{1/3} = d_s \left(\frac{\Delta \rho g}{\rho_f \nu^2} \right)^{1/3}, \qquad (2.8)$$

in u_* which does not arise. Based on their data collection Yalin and da Silva (2001) provide one explicit equation as an approximation, namely

$$\tau_c^* = Y_{cr} = 0.13 \Xi^{-0.392} e^{-0.015 \Xi^2} + 0.045(1 - e^{-0.068 \Xi}) \,, \qquad (2.9)$$

where the Yalin parameter Ξ is a dimensionless variable, which is equal to the dimensionless grain diameter D^* given in equation (2.8).

2 Literature Review

Empirical threshold curves where also assessed by Paphitis (2001), who considered 29 different data sets from the last century. The data cover the range $0.01 < \text{Re}^* < 10^5$. He provides different explicit expressions for combinations of dimensionless parameters τ_c^* versus Re^* or D^* as well as dimensional parameters τ_c or u_{*c} versus d_s. Below, only the traditional Shields relation is given, plus the lower and upper limits of the collected data.

mean threshold values:
$$\tau_c^* = \frac{0.188}{1 + \text{Re}^*} + 0.0475(1 - 0.699e^{-0.015\,\text{Re}^*}) \tag{2.10}$$

lower limit:
$$\tau_c^* = \frac{0.075}{0.5 + \text{Re}^*} + 0.03(1 - 0.699e^{-0.015\,\text{Re}^*}) \tag{2.11}$$

upper limit:
$$\tau_c^* = \frac{0.28}{1.2 + \text{Re}^*} + 0.075(1 - 0.699e^{-0.015\,\text{Re}^*}) \tag{2.12}$$

2.1.2.3 Stochastic or Probabilistic Approaches

In consideration of Fig. 2-3 the ambiguity in the determination of a value for the critical shear stress is obvious. Due to this and to account for the random nature of turbulence and sediment movement, some researchers developed approaches which describe and quantify an observable state of motion, rather than a hypothetical state of zero movement. These kinds of approaches are termed probabilistic or stochastic.

One of the first derivations of a stochastic concept for bed load transport was presented by Einstein (1937) within his doctoral thesis. Einstein (1950) defined the pickup probability for a particle as "the probability of the dynamic lift force on the particle being larger than its weight (under water)". For the evaluation of the pickup probability p_e, he derived theoretically the following formula:

$$p_e = 1 - \frac{1}{\sqrt{\pi}} \int_{-B_*\Psi_* - 1/\eta_0}^{+B_*\Psi_* - 1/\eta_0} e^{-t^2} dt \tag{2.13}$$

where $\Psi_* = 1/\tau_b^*$ is the flow intensity, $\eta_0 = 0.5$ is the standard deviation and t is the only variable of integration. The constant $B_* = 1/7$ was obtained for uniform sediment by using the data of Meyer-Peter et al. (1934) and others. The non-central probability density function (abbreviated as pdf) on the right hand side of equation (2.13), i.e. the definite integral of the Gaussian normalized by its total area $\sqrt{\pi}$, can be interpreted as the probability for a particle being stationary for a given flow intensity Ψ_*.

A solution for the integral in closed form in terms of elementary functions does not exist but can be gained by approximation. Cheng and Chiew (1998) reviewed approximations of different authors and present their own equation for the probability p as a function of the dimensionless shear stress τ_b^* and the lift coefficient C_L for hydraulically rough flow. Assuming a mean critical value $\tau_b^* = \tau_c^* = 0.05$ for the horizontal part ($\text{Re}_c^* \geq 500$) of the Shields curve, they arrive at a probability of about 0.6 % for $C_L = 0.25$. By their interpretation, this implies that 0.6 % of all the particles, on a given bed area, are in motion. A summary of appropriate values ranging from 0.008 % to 0.3 % obtained by models of other researchers is provided by Papanicolaou (1999), indicating over-prediction of the former approach. In comparison, the theory of Einstein (1950)

2.1 Modelling of Bed Load Transport

with $C_L = 0.178$ gives a value of approximately 13 % for the same τ_b^*. Furthermore, Gessler (1965), and subsequently Günter (1971), concluded from their deliberations on the effective shear stress that the values for the critical mean shear stress gained by extrapolation to zero bed load transport as applied by Shield or Meyer-Peter (e.g. $\tau_c^* = 0.047$) correspond to a state with approximately 50% particles in motion. This agrees with the conjecture of Meyer-Peter that for $\tau_c^* = 0.047$ bed load could be already expected. Gessler and Günter take up the position that the mean shear stress acting continuously on the particles on the bed is distinctly smaller than the local instantaneous shear stress due to turbulent fluctuations. It follows that particle motion or bed load transport can occur for a smaller effective shear stress than the critical one.

The remarkable differences in the results may be further explained by the different determinations of the probability. On the one hand Einstein tuned his approach to cover sediment transport rates of flume experiments available at that time. Unfortunately his fit is quite poor but also very sensitive to the flow intensity in the range of small bed load transport intensity. On the other hand the newer approaches mentioned in this thesis are based on statistical analyses of experimental data (e.g. Jain (1992)) resulting in explicit probabilities for acting instantaneous forces like lift or drag, leading to an isolated view not including the whole spectrum of the processes. Consequently, a better interpretation of the above result by Cheng and Chiew (1998) would be that 'for 0.6 % of all particles the critical lift force is exceeded', because the amount of particles in motion is obviously much larger.

Wu and Lin (2002) provide improved approximations for the probability distribution with application of a log-normal pdf for the instantaneous velocity at the bed and appropriate values for C_L best fitting experimental data. Further enhancements for smooth turbulent flows were contributed by Wu and Chou (2003), who, apart from lifting also considered rolling probabilities and defined the mean total probability of entrainment as the sum of both[2]. The results reveal that a distinct probability for the critical state of sediment entrainment cannot be found, i.e. a critical shear stress does not exist. This finding is also acknowledged by other researchers (see review by McEwan and Heald (2001)). A further refinement of the approaches noted above was presented by Wu and Yang (2004), incorporating near-bed coherent flow structures in terms of a higher-order Gram-Charlier type pdf of near-bed streamwise velocity. Their approach accounts for turbulent bursting and mixed-size sediment and is applicable for hydraulically smooth and rough turbulent flow conditions. Also Hofland and Battjes (2006) derived a pdf for instantaneous drag forces and shear stress in turbulent rough and smooth flow that is in good agreement with the measurements. Cheng (2004) provides a formula for the estimation of the erosion probability especially for laminar flow that depends on the dimensionless shear stress and dimensionless particle diameter or the grain Reynolds number (see Fig. 2-3).

The stochastic model of Papanicolaou *et al.* (2002) accounts for the role of near-bed turbulent structures and various bed packing conditions of uniformly-sized particles on a flat bed. Based on a theoretical derivation and on experimental measurements, they provide values for the probability of the first displacement of a single sphere, which is, according to their hypothesis, equal to the probability of the exceedance of strong turbulent episodic events dislodging a sphere. Subsequently,

---

[2] This implies mathematically that lifting and rolling are statistically independent processes.

2 Literature Review

Dancey et al. (2002) introduced a new criterion for the threshold of motion of uniformly sized spherical particles with regard to the sediment bed packing density. Their new criterion describes the threshold of motion by means of a dimensionless parameter, which they interpreted as the probability of individual grain movement. Furthermore, they declare that for the condition of inception of motion, a specific, low, but nonzero value of the probability of grain movement applies. The topic of stream forces acting on a particle and the influence of turbulence is discussed in detail in chapter 3.4.

2.1.2.4 Concluding Remarks

The fundamental different views of the two approaches described above for incipient motion - the conventional threshold criterion according to Shields and the stochastic or probabilistic approach - are depicted in Fig. 2-4. The first is most commonly applied in river engineering due to its simple use. However its correct application requires calibration and experience, especially because of the explicit form of the motion threshold. Besides Einstein's or comparable subsequent work, the latter is still subject to current research driven by new measurement techniques allowing for a more and more detailed insight into the flow properties. For both, useful bed load transport formulas exist. Selected transport relations are presented in the next chapter.

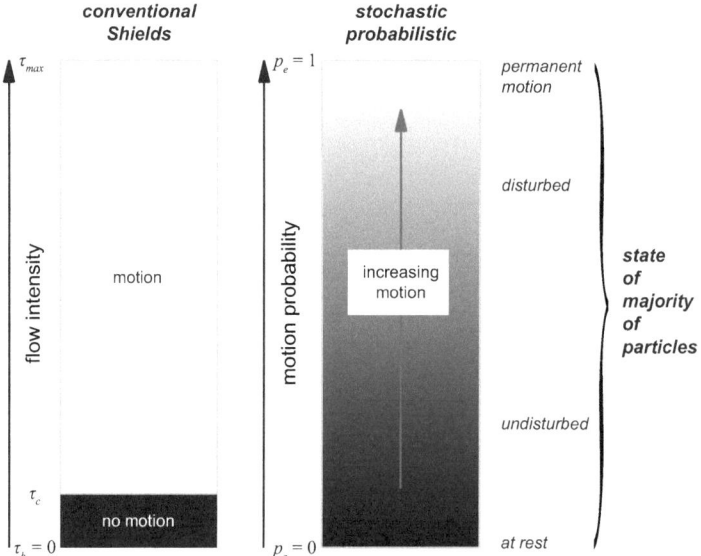

Fig. 2-4: Comparison of the two different concepts for particle motion

In addition, Fig. 2-4 shows a classification according to the state of the majority of the particles in direction of the flow intensity and the motion probability, respectively, which is valid for both models. In Bagnold's sense (Bagnold (1936)), the expression 'undisturbed' describes a particle which has not been displaced, whereas a particle that is displaced and then is resting is called 'disturbed'. Furthermore, the rather undefined state of permanent motion, where $\tau_b = \tau_{\max}$ and

$p_e = 1$, can be interpreted as transition from bed load to suspended transport, ending up in hyperconcentrated (e.g. Shu and Fei (2008)) or debris flow (e.g. Rickenmann (1999), Tognacca et al. (2000)).

2.1.3 Empirical Relations for Bed Load Transport

In engineering practice, a common approach to estimate sediment discharge in rivers is the application of empirical transport relations or sampling devices. The former are able to predict the transport rates only approximately and the latter require a large effort and may still not be very accurate. One of the first mechanistic relations was introduced by du Boys (1879). Subsequent flume experiments showed that du Boys' notion of the transport process as moving shearing layers of sediment to be fallacious (Ettema and Mutel (2004)). Nevertheless, the concept of a critical stream force exerted on the river bed causing sediment transport was appealing. Since then many researchers have contributed to the topic, resulting in a vast variety of bed load transport relations, but still no formulation can be claimed to be of universal applicability. Two basic kinds of transport relations have to be distinguished: total load formulas including suspended sediment load (e.g. Yang (2005) with emphasis on the Yellow River) and formulas considering exclusively bed load, which has been the focus in this chapter.

Graf (1971) provides a classification of bed load transport relations into three different types. One class defines du Boys-type equations which have a shear stress relationship describing the occurrence of bed load discharge only if the critical shear stress is exceeded, i.e. transport is a function of the residual shear stress $\tau_r = \tau_b - \tau_c$. The second class is made up of Schoklitsch-type equations (Schoklitsch (1926)) which are quite similar to those of the du Boys-type, but instead of shear stress, they are based on a flow discharge relationship, i.e. $q_r = q - q_c$ where q is the actual flow discharge, q_c is the water discharge at which the sediment begins to move and q_r is the residual discharge available for bed load transport. The third class are Einstein-type equations, based on statistical considerations of the lift forces. There are also other kinds of bed load transport formulas which are a combination of the above types or which relate the bed load discharge to a certain flow variable in terms of a power law with or without considering a threshold value. The latter may be interpreted as maximum idealisation of the relevant processes for sediment transport which provides an estimate for the bed load discharge.

An overview of the different types of bed load discharge formulas and their applicability to conditions of alpine rivers is presented by Habersack and Laronne (2002). Based on comparing sampling data of the gravel-bedded Drau River to calculated discharge they provide a ranking of the equations applied and suggest that a small, minimal number of bed load transport measurements should be undertaken in the field in order to check the empirical parameters used in the equations. Barry et al. (2007) provide a similar overview for gravel-bed rivers in the USA and present a general power law equation in terms of a flow discharge relationship to predict the bed load transport rates. Some examples of the different types will be discussed below.

Since the sediment in natural rivers consists of grains with almost continuous size, the sediment mixture has to be discretised to allow for discharge calculations. Hence, the corresponding grain size distribution is usually characterised by grain-size classes or by typical grain diameters, e.g. the mean diameter of the sediment mixture. This leads to another classification of bed load transport equations, namely relations for uniform and nonuniform sediment transport. Most of the available transport formulas were developed for uniform sediment because the complexity of the resulting relation could be significantly reduced and their application is quite handy. Furthermore, many of the bed load transport formulas have been derived from experimental flume data for near-equilibrium sediment transport and steady uniform flow conditions, e.g. Meyer-Peter and Müller (1948), Bagnold (1956), Camenen and Larson (2005) and many more.

The formula of Meyer-Peter and Müller (MPM) is currently widely used and was developed at VAW. It has its origin in the early correction works at the Alpine Rhine. The straightening of the channel led to a higher transport capacity but also caused aggradations in the lower river reach. This problem initiated an extensive laboratory study conducted at VAW under the guidance of Professor Meyer-Peter during a period of 16 years. The aim of that work was to provide a new channel design that is able to prevent further aggradations and ensures flood safety. To estimate the bed load transport for the given conditions and new channel geometries, the few existing formulas were far from being dependable or useful. Thus a better understanding of the fundamental processes and a new approach was needed. After the first series of laboratory flume experiments and field measurements in the Alpine Rhine by Meyer-Peter et al. (1934) in the early 1930ies, Meyer-Peter and Müller (1948) carried out additional investigations leading to their widely used MPM formula,

$$q_b = 8 \left[\frac{q'_w}{q_w} \left(\frac{k_b}{k_r} \right)^{3/2} \tau^* - 0.047 \right]^{3/2} \left((s-1) g d_m^3 \right)^{1/2} , \qquad (2.14)$$

with

$$\tau^* = \frac{\tau_b}{\triangle \rho g d_m} , \qquad (2.15)$$

where q_b is the volume bed load transport rate (volume flux of pure sediment without pore space per unit time) per unit channel width, k_r is the Manning-Strickler coefficient relative to the representative grain diameter of the mixture d_m, k_b is the Manning-Strickler coefficient including bed forms, q_w is the volume discharge of water per unit channel width with sidewall correction and q'_w that without any sidewall correction (see e.g. Wong and Parker (2006)). Moreover, $s = \rho_s / \rho_f$ is the sediment specific density. The experiments were performed for different slopes of 0.1‰ to 2.3% and with different uniform bed-material with grain diameters ranging from 5 mm to 28 mm. The formula is considered most appropriate for wide channels (for which $q'_w / q_w \to 1$) and coarse material. Meyer-Peter and Müller proposed in their original work a constant Shields parameter of $\tau^*_c = 0.047$ for fully turbulent flow ($\text{Re}^* > 70$). For smaller values of Re^*, the Shields parameter τ^*_c can be evaluated according to section 2.1.2.2 or a different transport formula may be applied. There exist various modifications and extensions of the MPM formula. For plane beds without bed forms, Wong and Parker (2006) present two simpler power law equations based on the reanalysis of the original MPM data, since the form drag correction is not necessary for such situations. Smart and Jäggi (1983) extended the MPM formula for channels and rivers with steeper slopes than those used in the MPM experiments. They studied bed load transport for slopes in the range from 3% to 20%. Another extended formula for the entire slope range, which can also be used for hyperconcentrated flows, was developed by Rickenmann (1991).

The application of uniform sediment transport relations to predict the bed load discharge is very common in engineering practice because they are quite simple and handy. However, the uniform approach is unable to account for sorting effects due to different grain sizes, so that uniform sediment transport formulas, such as the MPM formula, may over-predict the transport rate when applied to sediment mixtures as reported by Hunziker and Jaeggi (2002). For an initially flat bed which consists of a sediment mixture, the finer grains are eroded faster than the coarse ones. This leads to a stronger exposure of the coarse grains at the bed surface and the mobility of fine grains is reduced due to the shielding by coarser ones. To account for these rather complex processes occurring at the surface of a channel bed, Egiazaroff (1965) suggested the reduction of the critical shear stress by a so-called hiding factor and Ashida and Michiue (1971) proposed an evaluation of the hiding factor depending on the ratio of the individual grain diameter to the mean value of the sediment mixture at the surface. This approach may be used in combination with a uniform sediment transport formula applied to the individual grain diameters of the sediment mixture. A more sophisticated model for fractionwise sediment transport is presented by Hunziker (1995), where the sediment transport rate is calculated according to an extended MPM formula as described in Hunziker and Jaeggi (2002), which is also appropriate for poorly graded sediments. Other surface-based transport models for mixed-size sediment are e.g. Parker (1990) or Wu et al. (2000b) or Wilcock and Crowe (2003). Furthermore, fractional sediment transport may be predicted by stochastically based formulas, as originally proposed by Einstein (1950). More recent approaches are presented e.g. by Sun and Donahue (2000) or Kleinhans and van Rijn (2002).

Another classification of sediment transport equations corresponds to the basic experimental conditions under which the relations were developed. Most of the common sediment transport formulas are developed based on equilibrium sediment transport where the bed load discharge is equal to the sediment transport capacity and where the flow is steady and uniform. For an alluvial reach this means that the sediment input is equal to the sediment output over a sufficiently long time interval. However, during unsteady flow conditions like floods or due to the change in sediment supply, the sediment transport may not correspond to equilibrium conditions for a certain period. Non-equilibrium sediment transport also occurs for other unsteady processes, such as scouring as observed at hydraulic structures or due to channel contracting or a discontinuous bed, or they may be caused by a rapid change of bed topography. The shortcomings of an equilibrium transport model used under non-equilibrium conditions has been studied by Bell and Sutherland (1983). They observed that the transport rate shows a spatial variation because the flow requires a finite length of bed to erode sufficient sediment to achieve equilibrium transport capacity. A concept to account for this effect is to relate the time rate of change of the bed level to the difference between the local non-equilibrium bed load Q_b and the equilibrium transport rate Q_e, see e.g. Phillips and Sutherland (1989). This concept can be formulated in terms of the special rate of change of the effective bed load flux (see e.g. Bui and Rodi (2008)),

$$\frac{dQ_b}{dx} = -\frac{1}{L_s}\left(Q_b - Q_e\right), \tag{2.16}$$

where L_s is the so-called non-equilibrium adaption length. The equilibrium sediment transport rate Q_e is determined corresponding to an empirical relation as introduced above. Empirical formulas are also available for the adaption length L_s, see e.g. Van Rijn (1987) and Phillips and Sutherland

2 Literature Review

(1989). Equation (2.16) can be solved by a common analytic approach for nonhomogeneous ordinary differential equations. For the two-dimensional case usually a numerical approach is preferred. Applications of the non-equilibrium model show that changes in the bed level due to non-uniform flow can be better reproduced than by the equilibrium model, see e.g. Bui and Rutschmann (2010). However, the quality of the results depends on an appropriate choice of the non-equilibrium adaption length.

This section is concluded by a statement of Habersack and Laronne (2002), who point out that for Alpine gravel-bedded rivers fractional bed load discharge formulas perform well and the most promising are formulas capable of being used for both equilibrium and non-equilibrium or partial transport conditions that incorporate stochastic concepts.

2.1.4 Numerical Modelling of Sediment Transport in Rivers

A common approach for the modelling of bed load transport in research and engineering practice is to treat the water and the sediment as two immiscible phases, where the fluid and the sediment are considered as continua, as depicted in Fig. 2-5. This allows for the application of common numerical simulation models for fluid flow in combination with an empirically based approach for the determination of the sediment transport rate as outlined in the previous section 2.1.3.

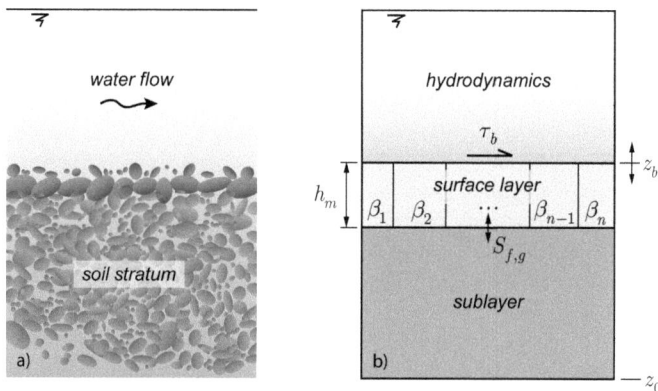

Fig. 2-5: a) vertical section in flow direction of a river where bed load transport occurs at the upper-most part of the river bed; b) common modelling approach for bed load transport; the water flow and the motion of the sediment are idealised by two separate continuum approaches.

Due to limited computational resources, one-dimensional models derived from the de Saint-Venant equations (see e.g. Cunge et al. (1980)), which use cross-sections for the discretisation of the channel or river reach, were preferred in the past. This approach is still very popular, especially when it comes to large-scale or long-term simulations. The two-dimensional de Saint-Venant equations are often termed shallow water equations. They use a horizontal mesh for the discretisation of the computational domain and the bed topography is usually based on a digital elevation model. Since the mentioned models are based on an integration of the three-dimensional Euler equations over the flow depth with the assumption of a hydrostatic pressure distribution, their applicability is restricted to flows where the streamlines are almost parallel and where the vertical component of the velocity can be neglected. Models using this approach are also termed "depth-averaged". Nevertheless, this simplification leads to governing flow equations where the flow depth or the water surface elevation is part of their solution; this is one of the reasons why they are so attractive for the modelling of free surface flows. However, their solution may be subject to numerical instabilities, which can be avoided by appropriate numerical techniques such as the introduction of numerical damping (see e.g. Chaudhry (2008)) or the application of Riemann solvers (see e.g. Toro (1997)). A corresponding simulation model which includes sediment transport and the coupling of the 1D and 2D models is proposed by Faeh et al. (2011). To account for density gradients, wind- and wave-driven currents or the curvature of streamlines, the two-dimensional shallow water equations can be extended to three-dimensions, still presupposing hydrostatic pressure conditions. Corresponding simulation models including sediment transport are proposed e.g. by Gessler et al. (1999) or Lesser et al. (2004).

With increasing performance of computer hardware and the use of parallelisation techniques, applications of three-dimensional models including sediment transport, which are based on the Navier-Stokes equations in combination with appropriate turbulence models, are now becoming popular. However, their application is limited to local scales and is only reasonable when three dimensional flow features play an important role. Turbulence models based on the Reynolds-Averaged Navier-Stokes equations (RANS), e.g. the k-ε model, are established in hydraulic engineering practice (see e.g. Rodi (1995)). For scientific research, also more sophisticated turbulence models such as the Detached Eddy Simulation (DES) (e.g. Spalart (2009)) or the Large Eddy Simulation (LES) (e.g. Rodi (2006)) are applied. Corresponding simulation models based on RANS, which include sediment transport and a k-ε turbulence model, are proposed e.g. by Wu et al. (2000a) or Olsen (2003). Spatially high-resolution models which include sediment transport are mainly used for research, e.g. the large-eddy simulation based model presented by Zedler and Street (2001). An approach using a meshfree method (SPH) for the simulation of the fluid flow in combination with a simple boundary shear stress erosion model and computed on the GPU (Graphics Processing Unit) is presented by Krištof et al. (2009).

The exerted force or rather the load due to the flowing fluid at the interface between the water and the sediment can be derived based on the properties of the flow. For depth-averaged hydrodynamic models, the boundary force may be expressed in terms of the bottom shear stress, which is obtained based on equation (2.3). To account for the various concepts for wall roughness, e.g. Manning-Strickler, Darcy-Weissbach or Chézy, the shear velocity u_* can be expressed in terms of the Chézy coefficient $c_f = \bar{u}_m / u_*$ and the mean or depth-averaged flow velocity $\bar{u}_m$. The resulting equation for the bottom shear stress reads

$$\vec{\tau}_b = \frac{\rho_f}{c_f^2} |\vec{\bar{u}}_m| \vec{\bar{u}}_m \ . \tag{2.17}$$

For the more advanced hydrodynamic models which include turbulent closure, the bottom shear stress may be obtained based on the turbulent quantities right at the boundary and/or within the turbulent boundary layer using a wall function.

2 Literature Review

To model sediment transport by a continuum approach, the soil stratum is horizontally discretised by cells and in the vertical direction by one or a few layers (see Fig. 2-5). If uniform sediment transport is considered, only one layer is necessary which corresponds to the whole soil stratum. For the sediment balance and the calculation of the temporal variation of the bed level z_b, equation (2.20) with $n_g = 1$ is solved, where n_g denotes the total number of grain size classes.

For non-uniform sediment transport with multiple grain size classes, two or more layers are used. To account for bed surface processes, bed load transport is considered to occur only in the surface layer and the subjacent layer, also called sublayer or active stratum, serves as sediment supply or deposit. This approach is also termed mixing layer or active layer concept, see e.g. Armanini (1995). As a distinctive feature, the surface layer is not fixed but vertically moves depending on erosion or deposition of sediment, which results in the change of the bed level z_b. Thus, the surface layer corresponds to a movable control volume for the bed load transport. To allow for bed material sorting, the conservation equation for sediment with constant density in one dimension,

$$(1-p)\frac{\partial(\beta_g h_m)}{\partial t} + \frac{\partial q_{b,g}}{\partial x} - S_{f,g} = 0 \quad for \ g = 1\ldots n_g, \tag{2.18}$$

is applied to the active layer, where p is the porosity of the sediment, β_g is the volume fraction of grain size class g, h_m is the height of the active layer, $q_{b,g}$ is the bed load transport rate of grain size class g per unit width and n_g is the total number of grain size classes. A similar equation holds for two-dimensional sediment transport. The source term $S_{f,g}$ describes the exchange of sediment with the active stratum due to the vertical dislocation of the active layer, i.e. the deposition of sediment from the active layer or the supply of sublayer sediment. The height h_m is constant or may be determined depending on the sediment mixture in the active layer (see e.g. Borah et al. (1982)). The exchange of sediment from the surface layer with the sublayer may also lead to a change of the sediment mixture in the active stratum. Thus, for the mass conservation in the sublayer with datum $z_0 = 0$ holds

$$(1-p)\frac{\partial(z_b - h_m)}{\partial t} + \sum_{g=1}^{n_g} S_{f,g} = 0. \tag{2.19}$$

The global conservation equation for the bed load, also termed Exner equation (Exner (1925)), relates the time rate of change of the bed level to the spatial variation of the volume rate of sediment transport per unit width. The corresponding control volume covers all sediment layers, i.e. the complete substrate of one cell. Hence, the global mass conservation reads:

$$(1-p)\frac{\partial z_b}{\partial t} + \sum_{g=1}^{n_g} \frac{\partial q_{b,g}}{\partial x} = 0. \tag{2.20}$$

A similar equation holds for two-dimensional sediment transport. The bed load flux $q_{b,g}$ may consist of several contributions, where the common bed load transport rate can be determined by an empirical approach as outlined in the previous section 2.1.3. Other contributions, such as transport due to transverse bed slope (see e.g. Ikeda (1982)) or gravitational transport due to slope collapse (see e.g. Volz *et al.* (2011)) may also be included. A numerical simulation model comprising the above approaches including suspended sediment transport is presented e.g. by Faeh *et al.* (2011).

An alternative to the active layer concept is proposed by Parker *et al.* (2000). They present a more general formulation of the erodible bed consisting in terms of a probabilistic Exner equation without the use of an active layer for mixed sized sediments. The temporal change of the bed level is described based on a probability density function of the bed elevation and elevation-specific sediment densities. However, this new approach requires additional data and a first closure for uniform sediment is presented by Elhakeem and Mran (2007). Another generalised form of the Exner equation is presented by Paola and Voller (2005), which includes e.g. soil formation, creep and compaction.

An overview comprising most of the above mentioned approaches and methods for hydrodynamics and sediment transport is given by Wu (2008). For the sake of completeness, the approach presented by Murray and Paola (1994) to model sediment transport in rivers has to be mentioned. They use quite a simple model based on cellular automata to simulate flow and sediment transport. The application of the model to the development of braided streams led to fairly realistic results. Based on this pioneering work, Thomas *et al.* (2007) developed a more advanced model. They used the cellular automata model to simulate the aggradations and degradations in braided river systems over a period of 200 years and obtained results consistent with those reported in the literature. Despite the rudimental hydraulics and the use of a simple power law for sediment transport, the results illustrate the potential of such models to simulate the behaviour of braided rivers for large spatial and temporal scales. However, when the morphological development depends on more complex flow features, such models may fail.

2.2 Numerical Methods and Investigations

2.2.1 Numerical Methods for Fluid Flow

Consideration of a fluid as a continuum is a common approach in physics and engineering. For the description of the dynamics of the fluid, caused by the change of internal and external quantities, appropriate conservation laws are postulated, where the quantities themselves are considered as infinitesimal parts of the continuum (see chapter 3). For the numerical solution of the conservation laws, the governing equations are transformed into a discretised form corresponding to an appropriate numerical method, where two basically different approaches have to be distinguished. From the Eulerian viewpoint, the time rates of change of the quantities are evaluated at spatially fixed points. The other approach is based on the Lagrangian viewpoint, where the discretisation points move depending on the flow. Notice that also the governing equations of the two approaches have a different form (see chapter 3.2.4.3). The first is the traditional approach for grid-based numerical methods for the solution of the governing equations of fluid dynamics. However, also grid-based methods with a combined Eulerian-Lagrangian approach exist, where the grid follows the deformation of the fluid. The latter is the usual approach for meshfree methods. Some common methods considering these approaches are subsequently outlined.

2.2.1.1 Grid-Based Methods

The Finite Difference Method (FDM) is the oldest method for the numerical solution of partial differential equations. For the application of the FDM, the differential form of the conservation equations is used, where the derivatives of the equations are replaced by ratios of finite differences in terms of the grid spacing. For the discretisation of the computational domain, structured grids are preferred, since these allow for the application of simple difference terms and the implementation of accurate higher order schemes. However, for the representation of arbitrary geometries special approaches are necessary. A detailed description of the FDM can be found in e.g. Roache (1998).

Another widely used method for computational fluid dynamics is the Finite Volume Method (FVM). For the approximation with the FVM the integral form of the conservation laws is used, thus the method is intrinsically conservative. The computational domain is subdivided into control volumes of polygonal shape, usually triangles or quadrilaterals, and the governing equations are applied to each control volume. Due to the unstructured grid, the FVM is suitable for the representation of complex geometries. An overview of FVM can be found in e.g. Eymard *et al.* (2000).

A similar method is the Finite Element Method (FEM), where also unstructured grids and discrete volumes in terms of finite elements are used. The main difference is that the equations are multiplied by a weight function before they are integrated over the domain. The main field of application of the FEM is structural mechanics and its application based on a Lagrangian description of the governing equations is popular. For its application to fluid dynamics see e.g. Glowinski (2003).

Another grid based method that has become more and more popular in recent years is the Lattice Boltzmann Method (LBM) (see e.g. Wolf-Gladrow (2000), Sukop and Thorne (2006)). This method is based on a simplification of the original concept of Boltzmann for the description of the dynamics of gas molecules. The governing equations for fluid flow are approximated by the use of a spatial lattice, or a grid, where the momentum exchange between the nodes is described by distribution functions. This approach has been successfully applied to various engineering problems primarily of fluid dynamics (see e.g. Mohamad (2011)).

One of the main difficulties which arise by the application of grid-based methods is due to the fact that a mesh is used for the simulation. Depending on the situation, the generation of an appropriate computational grid may be complicated and time consuming. Another challenge arises when movable or free boundaries have to be simulated. Both aspects are very common in hydraulic and river engineering. The hydraulic structures have a rather complex surface geometry and the flow in rivers is characterised by its free surface. In addition, when sediment transport is considered also the bottom becomes a movable boundary. For situations where local flow features are of interest, simplifications of the governing equation such as the shallow water equations are no longer applicable and a more complex flow equation such as the Euler or Navier-Stokes equations (see chapter 3.2.4) in combination with an appropriate treatment of the movable boundaries have to be solved.

Two basic approaches for the simulation of movable boundaries may be distinguished, namely *interface capturing* and *interface tracking*. In the first approach, the interface is captured based on a scalar marker quantity which changes according to the motion or deformation of the material, e.g. the fluid. This approach corresponds to an implicit formulation of the interface. A common interface capturing method for fluid interfaces of two-phase flows is the Volume-Of-Fluid (VOF) method (see e.g. Hirt and Nichols (1981), Rider and Kothe (1998)), where the marker quantity

2.2 Numerical Methods and Investigations

describes the volume-fraction of one fluid phase relating to a computational cell. Thus, the interface lies in cells where the volume-fraction is between 0 and 1. An approach used for fluid-solid interfaces, which is similar to the VOF method, is the Fractional-Area-Volume Obstacle Representation (FAVOR) method, which is also a fractional volume method; see Hirt (1993) for an overview on volume-fraction techniques. Another approach is the use of marker particles to identify different phases, e.g. fluid and solid, which corresponds to the Marker-And-Cell (MAC) method as proposed by Harlow and Welch (1965). Because the advection of the marker particles is computationally rather expensive, the application of this method is not very popular. However, due to increasing computing power, the method received new attention, see e.g. McKee *et al.* (2004) and Raad and Bidoae (2005). A more recent interface capturing method with application to many kinds of interface problems is the *level-set method* (see e.g. Adalsteinsson and Sethian (1995), Sethian and Smereka (2003)), where the marker quantity corresponds to a function describing the distance from the interface. For interface tracking methods, the deformation of the interface is explicitly described. Thus, the interface consists of discrete points. If these points are connected to the computational grids of the adjacent materials, the grids have to be adapted to follow the deformation of the materials. This approach is often used for the simulation of fluid-structure interaction problems in combination with an adaptive finite-element method (see e.g. Rannacher and Richter (2010)). Another approach where the surface points are not connected to the computational grids is the immersed boundary method (Peskin (1972)). This allows for the use of the above mentioned grid based discretisation technique, e.g. FDM, for the solution of the governing equations without the need of any special grid types (e.g. body-fitted) or grid adaption. However, the formulation of the boundary conditions in the vicinity of the interfaces is not straightforward. Mittal and Iaccarino (2005) give an overview on the subject and present applications to fluid-structure interaction problems.

Depending on the approach, the above methods can be used for the simulation of free surface flows or the modelling of the interaction of a fluid with a rigid or deformable solid. However, the numerical modelling of Fluid-Structure Interaction (FSI) is quite a challenging task and still an open topic with regard to the flexibility in the application of the approaches; see e.g. Bungartz *et al.* (2010). The main difficulty is to ensure that the applied temporal and spatial coupling techniques are energy-conserving and stable. For modelling of FSI, the pressure forces at the interface have to be passed to the deformable surface or the rigid body, and the corresponding response of the body has to be transferred to the fluid. Based on the preferred flow solver, the time integration may be explicit which may result in an unstable simulation or implicit to ensure stability. Furthermore, the accuracy and the level of detail of the acting fluid forces depend on the grid resolution and on the applied turbulence model. The application of eddy-resolving schemes to FSI is presented e.g. by Münsch and Breuer (2010).

2.2.1.2 Meshfree Methods

The term "meshfree methods", also called particle methods, corresponds to a wide field of numerical methods which are used either for the discretisation of a continuum or the representation of a discrete body; both kinds are used in the scope of this work. The first is discussed in the present section and the latter is outlined in section 2.2.2.

The basic difference between grid-based methods and meshfree methods is that no grid is necessary for the discretisation of the computational domain. For meshfree methods, a set of arbitrary distributed particles is used which represent the nodes required for the spatial discretisation. This permits to overcome many of the problems arising from the use of a computational mesh, especially the treatment of movable boundaries and the generation of grids as discussed in the previous section. Furthermore, meshfree methods seem to be a promising approach for the simulation of fluid-structure interaction as applied in this work. The approximation of the derivatives of the governing equations varies depending on the specific method, e.g. integral interpolation, finite differences or moving least squares, just to name a few. An overview on some common meshfree methods is given by Huerta *et al.* (2004), Nguyen *et al.* (2008) and Koumoutsakos (2005) shows the potential of particle methods for multi-scale flow simulations.

One of the first meshfree methods which also has a long continuous history is Smoothed Particle Hydrodynamics (SPH), originally presented by Lucy (1977) and Gingold and Monaghan (1977) for astrophysical simulations. With regard to hydraulic and environmental engineering, the method is well suited for the simulation of related problems. The application of the method to free surface flows is very common, see e.g. Monaghan (1994), Violeau and Issa (2007b), Crespo *et al.* (2007) and Fang *et al.* (2009). Lee *et al.* (2010) used the method for the simulation of the water collapse in waterworks and flow in a river dam spillway. The method was extended by Monaghan and Kocharyan (1995) for the simulation of multi-phase flows. SPH is also popular for the investigation of wave propagation and wave breaking, e.g. Monaghan and Kos (2000), Dalrymple and Rogers (2006) and Shao (2006). Various applications of SPH to fluid-structure interaction exist; the fluid motion due to the impact of a geometrical body distinctly larger than the particles has been simulated e.g. by Monaghan *et al.* (2003), Shao (2009) or by Ataie-Ashtiani and Shobeyri (2008) and Qiu (2008) in terms of landslide-generated waves. Recent overviews on SPH are presented e.g. by Gomez-Gesteira *et al.* (2010a) or Liu and Liu (2010). The SPH method is also popular in the field of computer graphics and animations, where the realistic representation of fluid flow or the behavior of bodies in the presence of a fluid is of interest (see e.g. Stam and Fiume (1995), Muller *et al.* (2003), Muller *et al.* (2004)). SPH is one of the methods applied in this work and a detailed description of the method can be found in chapter 4.2.

Another meshfree particle method which has been successfully applied to fluid dynamics are *vortex methods* (see e.g. Chorin and Bernard (1973), Leonard (1980)). The governing flow equations are based on the velocity-vorticity formulation of the Navier-Stokes equations, where the velocity field is obtained using a Poisson equation in combination with suitable boundary conditions (Cottet and Koumoutsakos (2000)). The particles used for the discretisation of the flow field represent so-called "blobs" of vorticity (Baker and Beale (2004)) and hence a vortex-blob approximation is used for the numerical interpolation. However, the formulation of appropriate solid boundary conditions (Koumoutsakos *et al.* (1994)) is not as straightforward as e.g. for FDM. The application of a vortex method in combination with a level set approach for the simulation of fluid-structure interaction is presented by Coquerelle and Cottet (2008).

For the sake of completeness, some combinations of Lagrangian und Eulerian approaches are briefly outlined. The Particle-In-Cell (PIC) method was developed in the middle of the last century (see e.g. Harlow (1964)) for the simulation of problems where a soil behaves like a fluid. For the PIC method, the deformation of the material is described by Lagrangian particles which carry mass and other information and an Eulerian mesh is used for the interpolation of the field variables. An improvement of the interpolation approaches used for PIC and the extension of PIC to SPH are presented by Monaghan (1985).

For FSI problems where detailed local forces may play an important role, meshfree particle methods are sometimes blamed to be of insufficient accuracy due to particle distortion. Thus, some methods propose a combination of grid-based and meshfree approaches with the aim to improve accuracy. Onate *et al.* (2004) presented the particle finite element method, where the nodes are freely movable particles and the governing equations are solved by a FEM based on a mesh constructed from the nodes. Corresponding applications to FSI problems are presented in Onate *et al.* (2008). Another approach termed remeshed SPH (rSPH), where SPH particles are remeshed by a moment conserving scheme to prevent the disordering of particles is presented by Chaniotis *et al.* (2002). The application of rSPH in combination with immersed boundary and level set techniques to self-propelled swimmers is presented by Hieber and Koumoutsakos (2008).

2.2.2 Modelling of Granular Material

Granular materials consist of a large number of individual grains of similar or nonuniform size. For the simulation of its dynamic behaviour, continuum or discrete approaches are used. One simple continuum approach in terms of sediment transport, where only changes concerning the surface of the granular material are considered and where the motion of the material is reduced to mass transport due to boundary shear stresses or gravitational forces, has been introduced in the previous chapter 2.1.4. To model more complex processes, where the granular material behaves like a fluid, e.g. as for debris flows or landslides, different approaches have to be applied. Reviews on the topic are given e.g. by Savage (1984) or Hutter and Rajagopal (1994). An appealing approach for this kind of flows is based on the shallow water equations, where the material is treated as a single phase with movable boundaries and where the material behaviour is described by constitutive relations (see e.g. Savage and Hutter (1989), Iverson (1997), Douady *et al.* (1999), Pudasaini and Hutter (2007)). These constitutive relations comprise the internal friction, which can be motivated e.g. from rheological experiments and so-called kinetic theories, which consider the interactions between the grains (see e.g. Hutter and Schneider (2010a), Hutter and Schneider (2010b)). Alternative approaches are presented e.g. by Snider *et al.* (1998), who use the equations for fluid dynamics in combination with a particle probability distribution function or by Cummins and Brackbill (2002), who applied the standard equations of continuum mechanics; for both approaches the numerical models are based on the PIC method.

A more natural way for the modelling of granular materials is to treat the granular material as a conglomerate of discrete particles or bodies. Unlike the continuum approach, the corresponding models consider the interaction between soft or hard particles based on detailed interaction forces and are based mainly on a Lagrangian description. An overview of such approaches is given e.g. by Herrmann and Luding (1998) and with special focus on computational aspects by Pöschel and Schwager (2005). One such approach is the Discrete Element Method (DEM) as applied in this work and described in detail in chapter 4.3.

The discrete or distinct element method was originally developed by Cundall in the early seventies for the analysis of rock mechanics problems and has been extended by Cundall and Strack (1979) for geotechnical modelling of granular soils idealised by discs or spheres. Its further extension to

2 Literature Review

polyhedral shaped particles is presented by Cundall (1988) and Hart et al. (1988). Since then, the DEM has served as an efficient numerical tool for solving many scientific and technological problems in various fields of engineering. An overview of applications in mechanical engineering is presented by Fleissner et al. (2007). Lanru and Ove (2007) present the application of DEM to rock engineering, and Tavarez and Plesha (2007) demonstrate the capabilities of the method for the modelling of solid materials. The simulation of material tests are presented e.g. by Kacianauskas et al. (2007) or by Gaugele et al. (2008). Simulations of industrial application include e.g. mixing processes in vessels (Bertrand et al. (2004)), ball mills (Powell et al. (2011)) and hoppers (Balevicius et al. (2011)), plug flow in pipes (Tsuji et al. (1992)) or fluidized beds (Kawaguchi et al. (1998)). In the scope of an uncertainty analysis for a particle model of granular chute flow, Fleissner et al. (2009b) used DEM for the simulation of landslides. Furthermore, Teufelsbauer et al. (2009) and Teufelsbauer et al. (2011) investigated the interaction between granular flow and rigid obstacles by application of DEM and experiments. Examples of FSI simulations based on DEM in combination with various approaches for fluid dynamics are presented in the subsequent section.

Cleary and Prakash (2004) point out the potential of discrete element modelling and SPH for applications in the environmental sciences. They show a variety of applications of both methods (not combined), e.g. to landslides, dam-breaks, tsunamis and volcanic lava flow.

2.2.3 Simulation of Fluid and Sediment Particles

In this section, some numerical investigations into the interaction of sediment particles and fluid flow are outlined. A brief overview on experimental investigations into the forces acting on particles with special focus on the processes at the channel bed is given in chapter 3.4.2.3.

Because of the complexity of the processes involved in the interaction of fluid and particles, corresponding numerical models are usually based on simplifications. A common approach is to use fixed particles, i.e. to study a porous medium such as filters. For movable particles, a reasonable simplification for some situations is to consider only the momentum exchange from the fluid to the particles and to determine the fluid forces acting on the particles by a simple drag law. Successful simulations based on such simplifications using a conventional Navier-Stokes solver were carried out; Tsuji et al. (1993) present the simulation of fluidized beds, Herrmann et al. (2007) for porous media and aeolian transport, where also the simulation of propagating sand dunes (Schwammle and Herrmann (2003)) has to be mentioned.

Another approach to model water-sediment mixtures is to consider the fluid and the sediment as two fluids with distinct properties which consist of particles. In contrast to the former approach, models of this kind include two-way coupling, i.e. the momentum exchange between the fluid and the sediment, and vice versa is considered. Applications based on this approach, where the sediment phase is represented by SPH particles, are presented e.g. by Monaghan et al. (1999), who studied gravity currents descending a ramp in a stratified fluid. Bui et al. (2007) simulated the soil-water interaction due to a vertical jet and examples with special focus on sediment transport are the simulation of waves generated by granular landslides by Falappi and Gallati (2007) or the modelling of rapid scour by Manenti et al. (2009). Numerical simulation of liquid-solid interaction in terms of a suspension using SPH is presented by Xiong et al. (2011). Shakibaeinia and Jin (2011) used this kind of approach to simulate the sediment transport due to a dam break, where a numerical method similar to SPH and constitutive laws for the sediment are applied.

In the last two decades, several investigations with regard to the interaction of sediment particles and fluid flow where the sediment particles are modelled by the discrete element method were carried out. Jiang and Haff (1993) used an experimental setup where the sediment bed consists of DEM particles and the fluid is modelled by a moving layer which exerts a velocity-dependent drag force on the embedded particles to study the micromechanics of bed load transport. A refined approach is presented by Schmeeckle and Nelson (2003). They studied three-dimensional bed load transport processes of mixed-size spheres and used a model similar to DEM to account for the interaction between particles. The exerted fluid forces on the spheres have been derived from the near-bed turbulent velocity field measured in laboratory experiments using detailed Laser Doppler Velocimetry (LDV).

Investigations into sediment transport based on the numerical simulation of the fluid flow in combination with the discrete element method for sediment particles were also accomplished. Cook *et al.* (2004) present a simulation method for a particle-fluid system which uses the lattice Boltzmann method to model fluid flow and an immersed boundary approach for the treatment of the interface. They show a two-dimensional simulation of the lateral erosion of elliptically shaped sediment grains at a flow contraction. They point out that despite the numerical efficiency of LBM and DEM (editor's note: at least for two-dimensional models), the simulation of large three-dimensional systems will be computationally expensive and a massively parallel implementation of the model is necessary. Feng *et al.* (2007) present a similar two-dimensional model which includes a LES model to account for small-scale turbulent effects. They show an application of the model to the sediment transport of spherical particles in a vertical pipe due to suction. Feng *et al.* (2010) extended this model to three-dimensions; the obtained simulation results are in good agreement with experimental data. A similar model using GPU acceleration is applied to the 3D flow past an array of spheres and to the settling of a multibody ensemble in a quiescent fluid by Owen *et al.* (2011). The direct numerical simulation, i.e. where all the details of the turbulent fluctuating fluid motion are resolved and thus no turbulence model is nescessary, of a settling sphere by application of the LBM combined with an approach for local adaptive grid refinement is presented by Yu and Fan (2010).

Last but not least, Potapov *et al.* (2001) present a strictly meshfree Lagrangian model based on a combination of DEM and SPH. They used the two-dimensional model to simulate the flow around a cylinder and the shear flow of particles between two parallel plates at low Reynolds numbers.

3 PHYSICS OF FLUIDS AND RIGID BODIES

3.1 Basic Description of a Conservation Law

For the derivation of a basic conservation law, consider an arbitrary spatially fixed domain Ω with boundary or surface Γ, surface normal $d\vec{\Gamma}$ and an infinitesimal quantity per unit volume Ψ (Fig. 3-1).

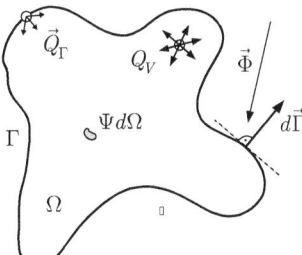

Fig. 3-1: Domain description for the derivation of a general conservation law

The variation per unit time of Ψ within the domain Ω should be equal to the contribution of incoming fluxes $\vec{\Phi}$ of quantity Ψ normal to the surface Γ plus contributions of boundary and volume sources of quantity Ψ, $\vec{Q}_\Gamma$ and Q_V. Consequently, the general form of the conservation equation for the quantity Ψ reads

$$\frac{\partial}{\partial t} \int_\Omega \Psi \, d\Omega = -\oint_\Gamma \vec{\Phi} \cdot d\vec{\Gamma} + \oint_\Gamma \vec{Q}_\Gamma \cdot d\vec{\Gamma} + \int_\Omega Q_V \, d\Omega \,. \quad (3.1)$$

With Gauss' theorem for continuous fluxes and surface sources[3] and for an arbitrary volume Ω, equation (3.1) can be written in the differential form

$$\frac{\partial \Psi}{\partial t} + \vec{\nabla} \cdot \vec{\Phi} = Q_V + \vec{\nabla} \cdot \vec{Q}_\Gamma \,. \quad (3.2)$$

---

[3] $\oint_\Gamma \vec{F} \cdot d\vec{\Gamma} = \int_\Omega \vec{\nabla} \cdot \vec{F} \, d\Omega$

3 Physics of Fluids and Rigid Bodies

The flux consists of two contributions: the convective flux $\vec{\Phi}_C = \Psi \vec{u}$ that is the amount of Ψ transported with the flow of velocity $\vec{u}$ and the diffusive flux $\vec{\Phi}_D$ due to molecular, thermal agitation.

If the quantity itself is a vector $\vec{\Psi}$, the flux vector and the boundary source vectors become tensors, $\mathbf{\Phi}$ and $\mathbf{Q}_\Gamma$, and the scalar volume source a vector $\vec{Q}_V$. Thereby the convective flux is $\mathbf{\Phi}_C = \vec{u} \otimes \vec{\Psi}$. Analogous to equation (3.1) the conservation law then reads

$$\frac{\partial}{\partial t} \int_\Omega \vec{\Psi} \, d\Omega = -\oint_\Gamma \mathbf{\Phi} \cdot d\vec{\Gamma} + \oint_\Gamma \mathbf{Q}_\Gamma \cdot d\vec{\Gamma} + \int_\Omega \vec{Q}_V \, d\Omega \;, \tag{3.3}$$

and equation (3.2) becomes

$$\frac{\partial \vec{\Psi}}{\partial t} + \vec{\nabla} \cdot \mathbf{\Phi} = \vec{Q}_V + \vec{\nabla} \cdot \mathbf{Q}_\Gamma \;. \tag{3.4}$$

Note that the two equations (3.1) and (3.3) are generally valid and are to be considered as the basic formulation of a conservation law. They remain valid even if discontinuities exist in the variation of the conserved quantity. On the contrary, the formulations (3.2) and (3.4) are only valid if continuity and/or differentiability of the properties can be assumed.

3.2 Governing Equations for Fluid Flow

3.2.1 Conservation of Mass

In Galileian mechanics, for a compressible fluid in a closed system, the total sum of mass will be constant since mass cannot be created nor disappear. In the absence of sources or sinks, the variation of mass over time is exclusively due to flux across the boundary. With regard to the formerly introduced conservation law, density is actually the quantity to be conserved, i.e. $\Psi = \rho_f$. Since no diffusive flux exists for mass transport, mass will only be transferred by convection. Thus the flux term for a fluid with velocity $\vec{u}$ reads $\vec{\Phi}_C = \rho_f \vec{u}$. The resulting conservation law of mass can be written as follows:

$$\frac{\partial}{\partial t} \int_\Omega \rho_f \, d\Omega = -\oint_\Gamma \rho_f \vec{u} \cdot d\vec{\Gamma} \;, \tag{3.5}$$

and the corresponding differential form reads

$$\frac{\partial \rho_f}{\partial t} + \vec{\nabla} \cdot \left(\rho_f \vec{u} \right) = 0 \quad or \quad \frac{D\rho_f}{Dt} + \rho_f \vec{\nabla} \cdot \vec{u} = 0 \;, \tag{3.6}$$

in which

$$\frac{D(.)}{Dt} := \frac{\partial(.)}{\partial t} + \vec{u} \cdot \vec{\nabla}(.) \tag{3.7}$$

is the substantive or material derivative.

3.2.2 Conservation of Momentum

The fundamental equations describing the motion of a viscous Newtonian fluid are the conservation equations of momentum. They are based on the work of Navier at the beginning of the 19th century. Investigations of Poisson, de Saint-Venant and Stokes at a time only a little later led to the same results. Hence, they are known as the Navier-Stokes equations.

The physical background of the momentum equations are Newton's laws of motion as introduced later in section 3.3.1. The motion of a fluid is caused by applied external volume forces $\vec{f}_e$ per unit mass, i.e. $\vec{f}_e = \vec{F}_e / (\rho_f d\Omega)$, and its deformation is related to internal stresses σ. The internal stresses have two contributions namely viscous stresses τ and pressure p

$$\sigma = -p\mathbf{I}_1 + \tau , \tag{3.8}$$

where $\mathbf{I}_1$ is the unit tensor. For a Newtonian fluid, the internal stresses can be related to the rate of strain. When applying a linear stress law that can be expressed in three dimensions by the stress tensor

$$\tau = \tau_{ij} = \mu\left(\partial_i u_j + \partial_j u_i\right) + \lambda\left(\vec{\nabla} \cdot \vec{u}\right)\delta_{ij} , \tag{3.9}$$

where μ is the dynamic viscosity of the fluid and $\lambda = -2\mu/3$ according to the Stokes relation. Since the internal stresses act as surface sources and the external forces as forces per unit volume $\rho_f \vec{f}_e$, the sum of the sources, i.e. the applied forces in the given case, can be written as:

$$\oint_\Gamma \sigma \cdot d\vec{\Gamma} + \int_\Omega \rho_f \vec{f}_e \, d\Omega . \tag{3.10}$$

The quantity to be conserved is momentum and thus $\vec{\Psi} = \rho_f \vec{u}$ (momentum per unit volume) and with conservation of mass, the flux only consists of its convective part. According to Newton's second law, the time rate of change of momentum has to be equal to the applied forces. By application of Gauss' theorem and by inserting (3.8) into (3.10) this leads to the following conservation equation:

$$\frac{\partial}{\partial t}\int_\Omega \rho_f \vec{u} \, d\Omega + \int_\Omega \vec{\nabla} \cdot \left(\rho_f \vec{u} \otimes \vec{u}\right) d\Omega = -\int_\Omega \vec{\nabla} p \, d\Omega + \int_\Omega \vec{\nabla} \cdot \tau \, d\Omega + \int_\Omega \rho_f \vec{f}_e \, d\Omega , \tag{3.11}$$

where the time rate of change of momentum is written on the left-hand side and the applied forces are on the right. Equation (3.11) can be written in non-conservative divergence form as

3 Physics of Fluids and Rigid Bodies

$$\rho_f \frac{\partial \vec{u}}{\partial t} + \rho_f \left(\vec{u} \cdot \vec{\nabla} \right) \vec{u} = -\vec{\nabla} p + \vec{\nabla} \cdot \boldsymbol{\tau} + \rho_f \vec{f_e} \tag{3.12}$$

and is called the Navier-Stokes equations for $\boldsymbol{\tau}$ according to equation (3.9). Note that an equivalent equation may be derived by combination of equations (3.27), (3.6) and (3.10). Detailed derivation of the Navier-Stokes equations can be found in any book on continuum mechanics and e.g. in Hirsch (1988), Schlichting *et al.* (2000) or Hutter and Jöhnk (2004).

3.2.3 Conservation of Energy

Since the Navier-Stokes equations, in their original form as above, are formulated for a compressible fluid like a gas (in which case $\lambda \neq -2\mu/3$), the state of the corresponding thermodynamic system has to be considered. For energy the fundamental law holds that it cannot be created nor disappear, it just can be transformed into another form of energy. The total energy per unit mass of a fluid is the sum of its internal energy e per unit mass and its kinetic energy per unit mass,

$$E = e + \frac{\vec{u}^2}{2} \ . \tag{3.13}$$

The internal energy is a state variable of the system that is related to temperature T. Hence, the time rate of change of total energy is due to the convective energy and momentum fluxes, the flux due to heat conduction, the power of internal forces and the power of external forces $P_f = \rho_f \vec{f_e} \cdot \vec{u}$ as well as sources q_H other than conduction (e.g. chemical reactions). This results in the conservation equation of energy that can be written in differential form as

$$\frac{\partial}{\partial t}\left(\rho_f E\right) + \vec{\nabla} \cdot \left(\rho_f \vec{u} E\right) = -\vec{\nabla} \cdot \vec{q_v} - \vec{\nabla} \cdot \left(p\vec{u}\right) + \vec{\nabla} \cdot \left(\boldsymbol{\tau} \cdot \vec{u}\right) + P_f + q_H \ , \tag{3.14}$$

where $\vec{q_v} = -k\vec{\nabla} T$ is the heat flux vector expressed in terms of Fourier's law and k is the thermal conductivity coefficient.

There are three equations, the continuity equation (3.6), the momentum equation (3.12) and the energy equation (3.14), but five unknowns: density ρ_f, velocity $\vec{u}$, pressure p, internal energy e and temperature T. The system can be closed by considering a simplification of a real gas, e.g. a perfect gas. Hence an equations of state, namely the ideal gas law, holds which relates p to ρ_f,

$$p = (\gamma_a - 1)\rho_f e \ , \tag{3.15}$$

where γ_a is the adiabatic index. Furthermore, for a perfect gas the internal energy can be related to temperature by $e = c_v T$, where the specific heat capacity c_v is constant. With these additional equations the governing equations of fluid flow are complete.

3.2.4 Simplifications

3.2.4.1 Incompressible Navier-Stokes Equations

For liquids, one important characteristic is their great resistance to compression and as far as fluid dynamics is concerned, this allows to regard them as being incompressible for most purposes with high accuracy. A fluid can be considered incompressible if its Mach number is $\text{Ma} \ll 1$. The Mach number is defined as $\text{Ma} = u_{ref}/c_s$, where u_{ref} is the reference flow velocity and c_s is the speed of sound of the fluid. For an ideal gas, the speed of sound is $c_s = (\gamma R T)^{1/2}$, where R is the specific gas constant. Typical values are: $c_s \approx 330$ m/s for air and $c_s \approx 1500$ m/s for water. See the discussion in Batchelor (2005) or Panton (2005) for the limits of applicability of the assumption of incompressibility.

If a fluid is incompressible there will be no or negligible variation of the density over the domain of interest. Thus $D\rho_f/Dt$ in equation (3.6) equals zero and only $\vec{\nabla} \cdot \vec{u}$ remains. The emerging differential equation then reads

$$\vec{\nabla} \cdot \vec{u} = 0 \ . \tag{3.16}$$

Furthermore, considering isothermal processes, the conservation law of energy can be neglected. By combination of equations (3.16) and (3.9) the expression for the stress tensor can be simplified. Thus, equation (3.12) can be rearranged which leads to the momentum conservation equation for incompressible flow

$$\frac{\partial \vec{u}}{\partial t} + \left(\vec{u} \cdot \vec{\nabla} \right) \vec{u} = -\frac{1}{\rho_f} \vec{\nabla} p + \nu \Delta \vec{u} + \vec{f}_e \ , \tag{3.17}$$

where ν is the kinematic viscosity and $\Delta = \vec{\nabla}^2$ is the Laplacian.

Since there are two equations, (3.16) and (3.17), for two unknowns, a straightforward solution of equations would be expected. However, the system of equations is nonlinear due to the convective term of the momentum equation. For some simplified one- and two-dimensional cases with appropriate initial and boundary conditions analytical solutions exist (see e.g. Schlichting *et al.* (2000)). However, the approximate solution of the equations by a numerical approach is most common.

3.2.4.2 Euler Equations

Another simplification, often applied, is to consider the fluid to be inviscid. The assumption of an inviscid fluid may be appropriate for convectively dominated flows with large Reynolds number (ratio of inertial force to viscous force) where laminar boundary layer effects do not have a significant influence. The corresponding equations are called Euler equations. Note that their general formulation is for a compressible fluid.

They consist of the conservation of mass and the conservation of momentum without the viscous term,

$$\frac{\partial \rho_f}{\partial t} + \vec{\nabla} \cdot \left(\rho_f \vec{u} \right) = 0 ,\tag{3.18}$$

$$\frac{\partial \left(\rho_f \vec{u} \right)}{\partial t} + \vec{\nabla} \cdot \left(\rho_f \vec{u} \otimes \vec{u} \right) = -\vec{\nabla} p + \rho_f \vec{f_e} .\tag{3.19}$$

Furthermore, heat conduction is neglected in the energy equation which leads to the following form of the conservation equation of energy,

$$\frac{\partial}{\partial t}\left(\rho_f E \right) + \vec{\nabla} \cdot \left(\rho_f \vec{u} E \right) = -\vec{\nabla} \cdot \left(p \vec{u} \right) + P_f .\tag{3.20}$$

Similar to the general equations for fluid flow, equations (3.18), (3.19) and (3.20) can be closed by an appropriate equation of state, e.g. equation (3.15).

3.2.4.3 Lagrangian Form of the Euler Equations

For the derivation of the conservation laws in the previous sections the time-dependent quantities, i.e. density, velocity and total energy, were considered as infinitesimal parts of a continuum. From the Eulerian viewpoint which is well-established in computational fluid dynamics, their time rate of change has to be evaluated at fixed points, e.g. at $\left(x_i\ y_j \right)$ as depicted in Fig. 3-2. Hence, the history of a quantity is limited to these points and it is generally not possible to track the path of a fluid particle.

An alternative viewpoint is the Lagrangian description; it can be regarded as a natural extension of particle mechanics. The fluid is considered to consist of material particles that move with the flow. Each particle is identified by its initial position $\vec{r}_{i,0}$ and a quantity carried with the particle is given in Lagrangian variables by

$$\Psi = \Psi_L \left(\vec{r}_{i,0}, t \right) .\tag{3.21}$$

The position $\vec{r}_i$ of a particle can be obtained by its path function or trajectory $\tilde{r}_i$ (Fig. 3-2b)

$$\vec{r}_i = \tilde{r}_i \left(\vec{r}_{i,0}, t \right) .\tag{3.22}$$

Based on the path function, the velocity, $\vec{u}_i = \partial \tilde{r}_i / \partial t$ and acceleration, $\vec{a}_i = \partial^2 \tilde{r}_i / \partial t^2$ of a particle can be defined.

3.2 Governing Equations for Fluid Flow

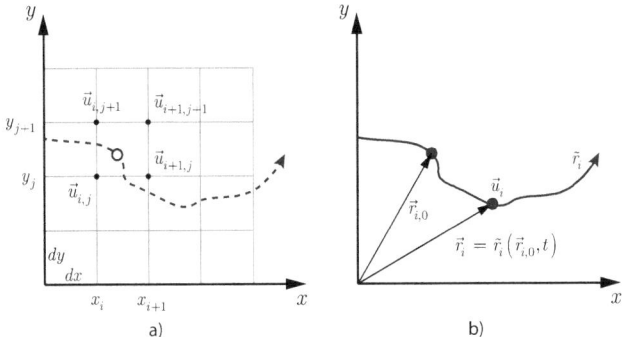

Fig. 3-2: Eulerian (a) and Lagrangian (b) viewpoint.

With this approach, the history of a particle can easily be tracked. However, since the Lagrangian analysis of fluid flow is usually quite difficult it is rarely applied (see e.g. Andrew (2005)). Nevertheless, when the fluid is discretised by particles as applied in this work, the use of the Lagrangian approach is reasonable (see chapter 4.2). Therefore, a time derivative for Eulerian variables is introduced that can be evaluated for a moving particle, called the substantive or material derivative (see definition equation (3.7)). By application of the substantial derivative, the Euler equations can be written in Lagrangian form as

$$\frac{D\rho_f}{Dt} = -\rho_f \vec{\nabla} \cdot \vec{u} \ , \qquad (3.23)$$

$$\frac{D\vec{u}}{Dt} = -\frac{1}{\rho_f}\vec{\nabla}p + \vec{f}_e \ . \qquad (3.24)$$

The energy conservation equation has been omitted, since for the present work the fluid is considered to be isothermal water at 20° C. Thus, the fluid is a liquid that generally can be regarded as incompressible. Nevertheless, under specific circumstances it may be necessary to take the small variation of density with change of pressure into account. Therefore, Batchelor (2005) presents an equation of state for water that is valid over a wide range of pressures,

$$\frac{p+B}{1+B} = \left(\frac{\rho_f}{\rho_0}\right)^{\gamma} , \qquad (3.25)$$

where ρ_0 is a reference density, B and γ are parameters[4]. For $B = 3000$ and $\gamma = 7$, equation (3.25) agrees with the physical properties of water to within a few per cent for pressures less than 10^{10} Pa. This relation is useful to obtain an approximate solution of the Euler equations as discussed in chapter 4.2.

---

[4] Fluids, whose equation of state is given by $p = p(\rho)$ are called barotropic.

3.3 Motion of Rigid Bodies

3.3.1 Equations of Motion

For moving bodies Newton's laws apply. The three laws describe the relation between the acting forces and the motion of the body.

Newton's first law states that a body with mass m at rest will stay at rest or the same body with velocity $\vec{v}$ will not change its velocity, if no unbalanced force acts on the body. The state of a body in motion can be described by its linear momentum as

$$\vec{p} = m\vec{v} \ . \tag{3.26}$$

Accordingly, the time rate of change of linear momentum, if not zero, demands an acting, non-balanced force $\vec{F}_a$. This fact is postulated by Newton's second law and reads (for the sake of consistency, Leibniz's notation will be used from now on)

$$\frac{d\vec{p}}{dt} = \frac{d(m\vec{v})}{dt} = m\frac{d\vec{v}}{dt} + \vec{v}\frac{dm}{dt} = \vec{F}_a \ . \tag{3.27}$$

Note that Newton's law generally applies for variable mass. The last term in the middle of the equation describes the time rate of change of mass which is of importance to describe the motion of a rocket for example. For constant mass Newton's second law reads

$$m\frac{d\vec{v}}{dt} = m\vec{a} = \vec{F}_a \ , \tag{3.28}$$

where $\vec{a}$ is the acceleration of the body. This special form is known as "mass times acceleration equals the sum of the forces".

Newton's third law describes the interaction of two bodies in contact; it is also called the law of action and reaction. Based on this third law, Newton derived the conservation of linear momentum that is elementary for the description of colliding bodies. In the absence of dissipative forces due to deformation, it states that the sum of linear momentum of the colliding bodies before and after collision is constant, i.e. linear momentum is conserved.

Newton's laws are said to deal with point masses; they describe the translational motion of an extended body only, while its rotation is not covered. Therefore, Euler introduced equations that describe the time rate of change of angular momentum; they are called Euler's equations (not to be confused with the homonymous equations for fluid dynamics from the same author discussed in chapter 4.2.3).

In analogy to equation (3.26) for linear momentum, angular momentum reads

3.3 Motion of Rigid Bodies

$$\vec{L} = \mathbf{I}\vec{\omega} \, , \tag{3.29}$$

where $\vec{\omega}$ is the angular velocity and $\mathbf{I} = \mathbf{I}^T$ is the tensor of moment of inertia. There is a Cartesian coordinate system (the fixed principal frame of the body) for which the inertia tensor is diagonal,

$$\mathbf{I} = \begin{bmatrix} I_x & 0 & 0 \\ 0 & I_y & 0 \\ 0 & 0 & I_z \end{bmatrix} . \tag{3.30}$$

For a rotationally symmetric body the inertia tensor can be further simplified, e.g. for a sphere the moment of inertia is $I_x = I_y = I_z = md_s^2/10$.

Similar to Newton's second law, the time rate of change of angular momentum is caused by the applied torque $\vec{M}_a$. Accordingly, the dynamic Euler equations read in the general vector form

$$\mathbf{I} \cdot \frac{d\vec{\omega}}{dt} + \vec{\omega} \times \left(\mathbf{I} \cdot \vec{\omega} \right) = \vec{M}_a \tag{3.31}$$

where $\mathbf{I} = \mathbf{I}^T$ is the tensor of moment of inertia in the fixed principal frame of the body.

Equations (3.28) and (3.31) are the equations of motion and actually the conservation laws for linear and angular momentum (for the basic description of a conservation law please refer to chapter 3.1.). They describe the time dependent motion of a body due to applied forces and torques and can be solved for their time dependent terms, i.e. the linear, $d\vec{v}/dt$, and angular, $d\vec{\omega}/dt$, accelerations.

3.3.2 Applied Forces and Torques

For a modelling approach like the discrete element method as used in this work, the forces $\vec{F}_{a,i}$ and contact torques $\vec{M}_{a,i}$ applied on a particle i are the sum of contact forces $\vec{F}_{c,ij}$ and torques $\vec{M}_{c,ij}$ due to interacting particles j plus external forces $\vec{F}_{e,i}$ or torques $\vec{M}_{e,i}$, respectively

$$\vec{F}_{a,i} = \sum_j \vec{F}_{c,ij} + \vec{F}_{e,i} \, , \tag{3.32}$$

$$\vec{M}_{a,i} = \sum_j \vec{M}_{c,ij} + \vec{M}_{e,i} \, . \tag{3.33}$$

The forces acting on a sphere surrounded by other spheres are depicted in Fig. 3-3.

3.3.2.1 Contact Forces
Contact forces are split into components normal and tangential to the contact surface and are treated differently depending on their orientation.

3 Physics of Fluids and Rigid Bodies

Normal forces:
The primary contact or interaction forces act normally to the contact surface. If the body is a sphere, they will only apply as a concentric force and thus not cause a torque at the centre. Several force laws used to model the interaction of rigid bodies in terms of spheres are discussed in chapter 4.3.2.

Tangential forces:
The secondary contact forces act tangentially to the contact surface and are due to friction. In most cases they lead to a torque. Two kinds of friction are distinguished, namely static and kinetic friction. In a static system, tangential forces due to static friction may be of importance, e.g. for a block on an inclined ramp. This effect can also be observed at sand piles. Kinetic or slip friction occurs when bodies interact with relative lateral velocities. Kinetic friction depends on the material properties of the interacting bodies and the normal force acting between them. Friction forces as used in this work are discussed in detail in chapter 4.3.4.

3.3.2.2 External Forces

The main external force is due to gravity. Assuming a constant acceleration of gravity of $\vec{g} = (0, 0, -g)$ with $g = 9.81$ [m/s²], the weight of a rigid body is given by

$$\vec{F}_g = V\rho\vec{g} = m\vec{g} \; , \tag{3.34}$$

where m is the mass of the body, V is its volume and ρ its density; for a sphere $V = d_s^3 \pi / 6$ with diameter d_s.
Other external forces and torques may be defined based upon the scope of the model, i.e. external forces and torques by virtue of initial and boundary conditions.

3.3.3 State of a Rigid Body

The translational state of a rigid body or particle i can be described by its position vector $\vec{r}_i$ and its velocity $\vec{v}_i$. According to the fundamentals of mechanics, these variables are related by

$$\frac{d\vec{r}_i}{dt} = \vec{v}_i \; ; \quad \frac{d\vec{v}_i}{dt} = \frac{d^2\vec{r}_i}{dt^2} = \vec{a}_i \; . \tag{3.35}$$

The rotational state of a particle can be described in a similar but more complicated way. This is because the three-dimensional Euler equations are generally nonlinear. The counterpart of the position of a particle is its orientation that can be described by its rotation unit quaternion $\mathbf{q}_i = \begin{bmatrix} q_{0,i} & q_{1,i} & q_{2,i} & q_{3,i} \end{bmatrix}$, i.e. $|\mathbf{q}_i| = 1$, in three dimensions. Furthermore, its angular velocity is described by $\vec{\omega}_i$. The time derivative of the angular velocity $d\vec{\omega}/dt$ can be obtained directly from equation (3.31) (see e.g. Fleissner (2010)). The time derivative of the quaternion can be obtained by

$$\frac{d\mathbf{q}_i}{dt} = \frac{1}{2}\mathbf{Q}(\mathbf{q}_i) \cdot \mathbf{\Omega}_i \; , \tag{3.36}$$

where $\mathbf{Q}(\mathbf{q}_i)$ is the orthogonal quaternion matrix and $\mathbf{\Omega}_i = \begin{bmatrix} 0 & \vec{\omega}_i \end{bmatrix}$ is a pure quaternion of the angular velocity (see Omelyan (1998)). Consequently, the second time derivative can be written as

$$\frac{d^2\mathbf{q}_i}{dt^2} = \frac{1}{2}\left(\mathbf{Q}\left(\frac{d\mathbf{q}_i}{dt}\right) \cdot \mathbf{\Omega}_i + \mathbf{Q}(\mathbf{q}_i) \cdot \frac{d\mathbf{\Omega}_i}{dt} \right). \tag{3.37}$$

Further details on the solution of the Euler equations using quaternions and quaternion calculus are given by Shabana (2010) or Vince (2008).

3.4 Fluid and Rigid Bodies

Two effects are crucial for a correct simulation of the different transport modes of rigid bodies in fluids. On the one hand, applied forces and torques on spheres occur due to their interaction, as introduced in the previous chapter (see Fig. 2-1). On the other hand, the applied forces and torques due to the presence of a fluid like water and its flow also play an important role. By way of illustration, both effects are composed in Fig. 3-3. In this chapter the basic hydro-mechanical forces are presented from an integral and partially empirical point of view. The approaches for the modelling of the detailed particle interaction forces are introduced in chapter 4.4. Besides general considerations of the interaction of a fluid with rigid bodies such as spheres, the relevant case of channel flow will briefly be discussed here.

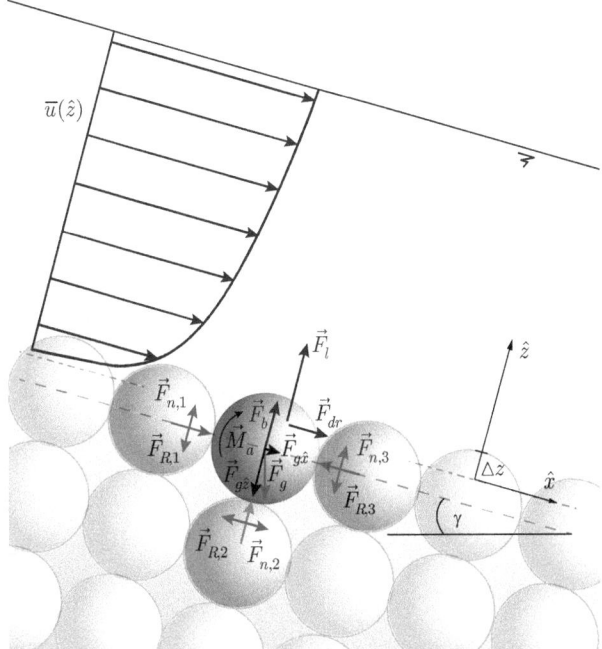

Fig. 3-3: Acting forces at a river bed consisting of spheres.

3 Physics of Fluids and Rigid Bodies

For channels with an inclined bed in the direction of the flow as depicted in Fig. 3-3, the slope is defined as $S_b = \tan\gamma$. In this work, the considered channel slopes are smaller than 0.01 which corresponds to alluvial rivers in Switzerland. This allows for simplification since for small values of γ, $\sin\gamma \approx \tan\gamma$ and $\cos\gamma \approx 1$. According to vector projection (see appendix A.5) the unit vectors tangential (streamwise) and normal (upwards) to the channel bed can now be written as $\vec{e}_{\hat{x}} \approx (1, 0, -S_b)$ and $\vec{e}_{\hat{z}} \approx (S_b, 0, 1)$. Furthermore, the components of the weight $\vec{F}_g$ in the $\hat{x}$- and $\hat{z}$-directions read

$$\vec{F}_{g\hat{x}} \approx m_s g \begin{pmatrix} S_b \\ 0 \\ -S_b^2 \end{pmatrix}, \left|\vec{F}_{g\hat{x}}\right| \approx m_s g S_b; \quad \vec{F}_{g\hat{z}} \approx m_s g \begin{pmatrix} -S_b \\ 0 \\ -1 \end{pmatrix}, \left|\vec{F}_{g\hat{z}}\right| \approx m_s g \,. \tag{3.38}$$

3.4.1 Hydrostatic Pressure and Buoyancy

Pressure is a fundamental property of a fluid that corresponds to a force $\vec{F}$ exerted on the surface A of an immersed body or a boundary by the fluid, i.e. $p = |\vec{F}|/A$ with units [N/m²]. Since pressure is a scalar quantity, its effect is independent of direction and it always acts in the normal direction to the boundary of the fluid.

Considering a static fluid of density ρ_f with a free surface and taking the atmospheric pressure as a datum, i.e. $p_{atm} = 0$, the pressure at an arbitrary point in the fluid depends on the water depth $h(z)$ at that point and is called hydrostatic pressure $p = \rho_f g h(z)$. The effect of hydrostatic pressure on a half sphere (sphere set onto the bottom of the channel) is schematically depicted in Fig. 3-4a. The resultant force $\vec{F}_h$ on a surface due to hydrostatic pressure is the sum of the forces on all surface elements of area dA

$$\left|\vec{F}_h\right| = \rho_f g \int_A h(z) dA = \left|\vec{F}_{hx} + \vec{F}_{hy} + \vec{F}_{hz}\right| . \tag{3.39}$$

The partition of the resultant force into its components often allows for simplification of geometrical bodies by the use of vector projection. In cases when the bottom is parallel to the water surface, the direction of the resultant hydrostatic pressure force is normal to the water surface. For the depicted half sphere with diameter d_s, the horizontal forces vanish, $\vec{F}_{hx} = \vec{F}_{hy} = \vec{0}$, because of symmetry and the vertical component is equal to the weight of the fluid on top of the sphere

$$\vec{F}_h = \vec{F}_{hz} = -\rho_f g \frac{d_s^2 \pi}{4}\left(h_f - \frac{d_s}{3}\right)\vec{e}_{\hat{z}} , \tag{3.40}$$

where h_f is the maximum water depth referring to the wetted surface of the sphere. The resultant force acting on the sphere in Fig. 3-4a) is

$$\vec{F}_{gh} = \vec{F}_g + \vec{F}_h . \tag{3.41}$$

3.4 Fluid and Rigid Bodies

The effect of buoyancy, also known as Archimedes' principle, occurs when a body has surfaces in contact with the fluid which have a normal in downward direction as depicted in Fig. 3-4b) for example. The buoyancy force is due to the pressure difference above and below the immersed body and is equivalent to the weight of the fluid displaced by the body

$$\vec{F}_b = \rho_f g V_s \vec{e}_{\hat{z}} \, , \tag{3.42}$$

where V_s is the volume of the body. The buoyancy force will act through the centroid of the volume of the displaced fluid which is equal to the centre of gravity if the body consists of homogenous material.

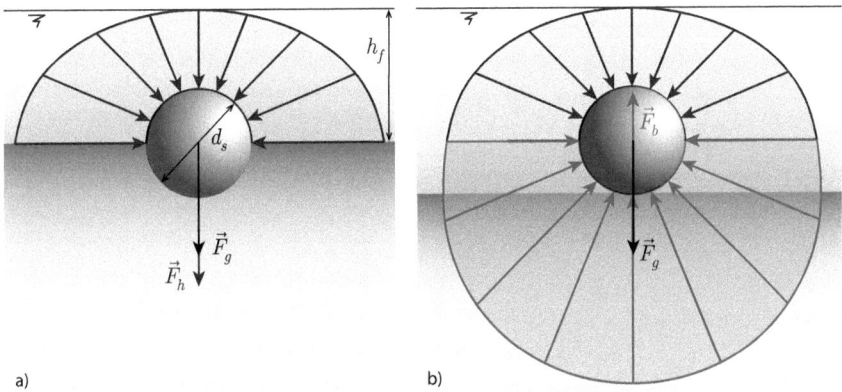

Fig. 3-4: Effect of hydrostatic pressure and resulting buoyancy.

With regard to the hydrostatic equilibrium of an immersed body, the buoyancy force can be seen as a reduction of its weight and is called submerged weight of the body

$$\vec{F}_{g*} = \vec{F}_{g\hat{z}} + \vec{F}_b = -\left(\rho_s - \rho_f\right) g V_s \vec{e}_{\hat{z}} \, , \tag{3.43}$$

where $\vec{F}_{g\hat{z}} = \vec{F}_g$ for a plane channel bed. Furthermore, the difference between the forces acting on the sphere for the two situations depicted in Fig. 3-4 is

$$\left|\vec{F}_{gh} - \vec{F}_{g*}\right| = \left|\vec{F}_h - \vec{F}_b\right| = \rho_f g \frac{\pi d_s^2 \left(d_s + 3h_f\right)}{12} \, , \tag{3.44}$$

and the ratio between them is

$$\frac{\vec{F}_{gh}}{\vec{F}_{g*}} = \frac{3\rho_f h_f + d_s \left(2\rho_s - \rho_f\right)}{2 d_s \left(\rho_s - \rho_f\right)} \, . \tag{3.45}$$

39

For example: for a sphere with $d_s = 0.03$ m, $\rho_s = 2800$ kg/m^3 and $h_f = 0.5$ m, $\rho_f = 1000$ kg/m^3, the ratio $\vec{F}_{gh}/\vec{F}_{g*} \approx 15$. Thus, use of an appropriate model allowing for fluid between the spheres of the channel bed (as illustrated in Fig. 3-3) and including buoyancy effects correctly may be crucial in order to obtain reliable simulations of bed load transport. This is also affirmed by the fact that the hydrostatic pressure distribution is implicit in the definition of the Shields parameter. The effect of non-hydrostatic pressure distributions on bed load transport has been pointed out by Francalanci et al. (2008) and is discussed in the next section based on occurring forces.

3.4.2 Hydrodynamic Forces

3.4.2.1 Drag of a Single Sphere

The force acting in the direction of relative motion of a body immersed in a fluid is called drag or fluid resistance. In other words, drag is the force in the direction of relative motion that has to be applied to move a body through a stagnant fluid or to keep the same body at rest in case of fluid flow (the weight of the body is neglected). The drag is made up of two contributions, namely the pressure drag arising from the non-uniform pressure distribution on the body and the skin friction drag due to shear stresses on the body surface. In a general form (see e.g. Douglas et al. (2001), the drag force is given as

$$\vec{F}_{dr} = C_D A_\perp \rho_f \frac{|\vec{u}_f|^2}{2} \frac{\vec{u}_f}{|\vec{u}_f|}, \qquad (3.46)$$

where $A_\perp$ is the area obtained by projection of the body on a plane perpendicular to the flow direction (i.e. $A_\perp = r_s^2 \pi$ for a sphere), C_D is the drag coefficient and $\vec{u}_f$ the free stream velocity (i.e. the uniform flow velocity upwind of the body). Note that $\rho_f |\vec{u}_f|^2 / 2$ is reminiscent the hydrodynamic pressure term of the Bernoulli equation.

A well-known solution for the drag of a sphere in steady uniform flow is that by Stokes, who obtained $\vec{F}_{dr} = 6\pi\mu r_s \vec{u}_f$ by simplification of the Navier-Stokes equations. Inserting this into (3.46) leads to a drag coefficient of $C_D = 24/\text{Re}$. However, Stokes' expression is restricted to laminar flow with $\text{Re} < 0.2$, also known as Stokes flow.

For higher Reynolds numbers, Karamanev (1996) states that one of the best relations to determine the drag coefficient is that by Turton and Levenspiel (1986) as it exhibits high precision and accuracy. Furthermore, he proposes a simplification of the original equation by introducing the Archimedes number Ar leading to a linear equation for the estimation of the final settling velocity (see chapter 1.1). Accordingly, the drag coefficient can be obtained in the form

$$C_D = \frac{432}{\text{Ar}}\left(1 + 0.047 \text{Ar}^{2/3}\right) + \frac{0.517}{1 + 154 \text{Ar}^{-1/3}}, \qquad (3.47)$$

where the Archimedes number for solid spheres is given by

$$\mathrm{Ar} = d_s^3 g |\Delta\rho| \rho_f / \mu^2 ~, \tag{3.48}$$

where $\Delta\rho = \rho_s - \rho_f$ and μ is the dynamic viscosity. For the regime of Newtonian flow, i.e. $500 < \mathrm{Re} < 10^5$, the drag coefficient for a sphere is practically constant and has the value $C_D = 0.44$.

3.4.2.2 Lift Force

Besides the drag force, the second component of the force exerted on an immersed body due to fluid flow is the lift force acting perpendicular to the direction of relative motion. It has the same origin as the drag force and can be defined in a similar way

$$\left|\vec{F}_l\right| = C_L A_\parallel \rho_f \frac{|\vec{u}_f|^2}{2} ~, \tag{3.49}$$

where C_L is the lift coefficient and $A_\parallel$ is the projected area for lift. In contrast to an airfoil in plane motion, where values for C_L depend on the angle of attack, the lift force for a sphere vanishes when $\vec{u}_f$ is uniform.

3.4.2.3 Forces Acting at the Channel Bed

For the flow over a channel bed made of spheres, as depicted in Fig. 3-3, the drag and lift forces may have different forms. Close to the rough boundary, the velocity is not uniform and the flow is turbulent. Since the acting pressure is a combination of hydrostatic and hydrodynamic pressure, the lift force may be reduced to a pressure difference that occurs due to the turbulent effects on the side of the sphere facing the flow; which has been measured by Einstein and Elsamni (1949) for hemispheres, by Dwivedi et al. (2010) for spheres, by Detert et al. (2010) for spherical as well as mixed sediments and by Smart and Habersack (2007) for natural gravel in a river, for example. Nevertheless, to allow for the application of equation (3.49) values for C_L are provided based on a stochastic analysis of experimental data (see e.g. Wu and Lin (2002)), as already introduced in chapter 2.1.2.3.

Investigations into the drag force exerted on a sphere set on top of a bed of closely packed spheres have been carried out by Coleman (1972). He concludes that the drag coefficient function for this situation corresponds with the function for a sphere in free fall. Schmeeckle et al. (2007) studied the situation of a sphere surrounded by other spheres without contact and for different exposure of the sphere to the flow. They obtained a drag coefficient of $C_D = 0.76$ which is independent of flow variability or particle exposure and is in agreement with the results of other studies. For decreasing exposure the drag force also decreases due to sheltering by the other particles while the lift force increases. The residual drag which exerts forces on the sphere can cause angular momentum; this is not covered in the mentioned study, but may be of importance in the process of particle entrainment.

For open channel flow, the most common engineering approach is to express the forces close to the bed by temporally and spatially averaged quantities, i.e. the bottom shear stress τ_b or the shear velocity u_*, as introduced in section 2.1.2.1. As a considerable generalisation the law for wall-

3 Physics of Fluids and Rigid Bodies

bounded turbulent flows holds and can be written for rough channel flow by considering Nikuradse's original parameterisation with $\kappa = 0.4$ as

$$\frac{\overline{u}(z)}{u_*} = \frac{1}{\kappa}\ln\frac{z}{k_s} + 8.48 = \frac{1}{\kappa}\ln\frac{29.7z}{k_s} \;. \tag{3.50}$$

For a velocity profile according to equation (3.50), the velocity is zero at $z_0 = k_s/29.7$. The distribution of the shear stress τ over the flow depth for the two dimensional case is given by (see e.g. Yalin and da Silva (2001))

$$\tau(z) = \tau_b\left(1 - \frac{z}{h_f}\right). \tag{3.51}$$

Equations (3.50) and (3.51) are depicted in Fig. 3-5. The offset of the frame of reference in the z-direction related to the tangential plane through the tips of the roughness elements is Δz. According to Bezzola (2002) $\Delta z \approx 0.25 d_s$ for uniform grains.

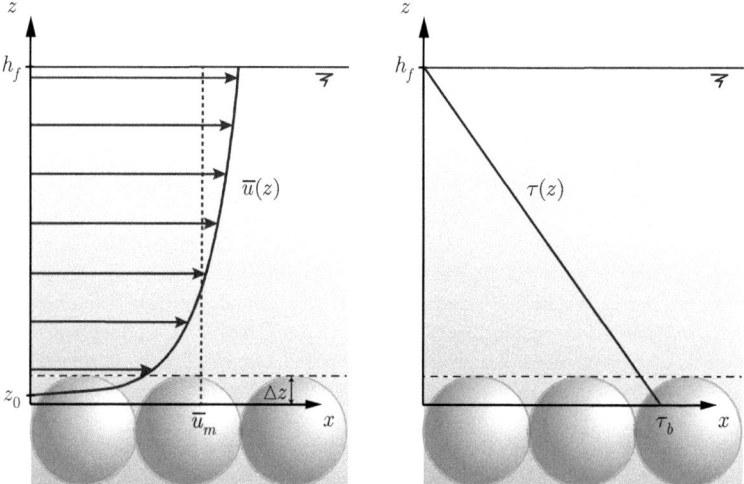

Fig. 3-5: Logarithmic law for turbulent channel flow

The shear stress expressed by the slope of the energy grade line or by the shear velocity as in equation (2.3) is an integral quantity that describes the flow resistance of the channel bed. Thus, it acts in opposite flow direction tangential to the bed. However, the quantities which entrain and move sediment are neither bed shear stress nor any other average characteristic of the flow, but instead the fluctuating forces, such as lift and drag, exerted directly by the flow on the particles, as stated by Schmeeckle et al. (2007). Thus, a force expressed in terms of $F = \tau_b A$ is a rough simplified model of reality.

Nonetheless, some researchers related their measurements of fluctuations of local forces or pressure to mean flow quantities. For example, Apperley and Raudkivi (1989) expressed the drag force in terms of the shear velocity as $\left|\vec{F}_{dr}\right| = C_* \rho_f u_*^2$, where C_* is a constant of proportionality relating the drag force to the shear velocity. By further analysis of their results, Schmeeckle et al. (2007) obtained for a sphere exposed to the flow the relation $\left|\vec{F}_{dr}\right| = C_D A_\perp \rho_f \, \overline{u}(d_s)^2 / 2$ where $C_D = 0.68$ and $\overline{u}(d_s)$ is the average flow velocity one particle diameter above the bed. Another approach that relates the shear stress to an indicator σ_p of the fluctuating lift is reported by Detert et al. (2010) and reads $\sigma_p / \tau_b = 2.88 \exp\left(2z/k_s\right)$.

The hydrodynamic forces discussed herein certainly play an important role for the incipient motion or entrainment of sediment particles, but peak values of the forces may not be sufficient. The duration of the peak values is also a significant factor as pointed out by Valyrakis et al. (2010). Therefore, they conjectured that impulse, rather than just the magnitude of hydrodynamic forcing, is relevant to the description of the incipient motion phenomenon.

4 NUMERICAL METHODS

4.1 Introduction

4.1.1 Methodology

The modelling approach applied in this thesis comprises the representation of the gravel bed and the water flow by particles which interact with each other. Therefore, Lagrangian methods, also called meshfree or particle methods, are applied to both the hydrodynamics and the bed load transport, which allows for a homogeneous discretisation of the underlying equations of motion. In other words, for discretisation of the computational domain and the multi-phase system, basically the same kind of approach is used; however with respect to the distinct properties of each phase, different methods are applied. The single grains of the gravel bed are modelled by discrete elements in the form of rigid spheres and their motion and interactions are resolved by application of the DEM. For the water flow, i.e. the hydrodynamic equations, the SPH method is applied. The modelling approach used is depicted in Fig. 4-1.

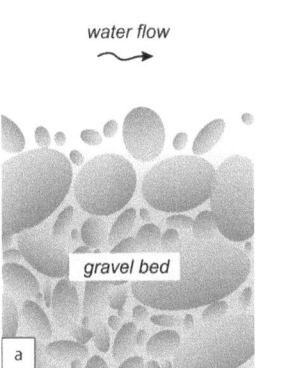 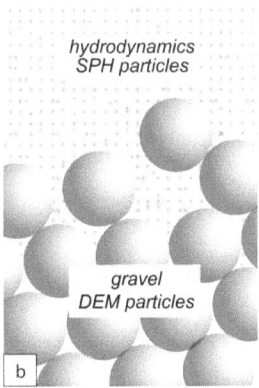

Fig. 4-1: Representation of (a) water flow and gravel bed by (b) SPH and DEM particles, respectively.

The primary advantage of this approach relies on the fact that any phase interface or fluid-structure coupling as well as interaction between solid objects is treated on a particle to particle basis, which makes the application of complex and expensive numerical methods as they are used for grid-based Eulerian approaches obsolete.

4.1.2 Software Framework *Pasimodo*

The numerical software used in this work is called *Pasimodo* ("Particle simulation and molecular dynamics in an object oriented fashion") which is a multi-purpose particle simulation tool developed by Fleissner (2010) at the Institute of Engineering and Computational Mechanics of the University of Stuttgart. The part for fluid dynamics simulation with SPH was mainly contributed by Lehnart (2008). The software is subject to continuous improvement and is applied by several academic collaborators.

The original purpose of *Pasimodo* was to simulate granular media in a dynamic environment. However, over the years, the software was extended by many features to allow for applications in a very wide reach of engineering problems. Besides granular media, the software can be used to model the dynamics and interaction of arbitrary shaped bodies, plastic-elastic rods, multi-body systems and fluids just to name a few of its capabilities. Some recent applications are: simulation of the cutting process or behaviour of an elastic membrane (Fleissner *et al.* (2007)), simulation of ductile cohesive material like aluminium (Gaugele *et al.* (2008)), tumbling sieving to separate different sized granular materials, such as sediments used in laboratory flume experiments, into different fractions (Alkhaldi *et al.* (2008)), landslides and granular chute flow in alpine regions (Fleissner *et al.* (2009b)), cargo sloshing in silo and tank vehicles (Fleissner *et al.* (2009a), Fleissner *et al.* (2010)).

For the definition and configuration of simulations *Pasimodo* uses XML input files that can be parameterized, also by time-dependent variables. Furthermore, the software is programmed in C++ following a transparent object-oriented design. The software can be extended or adapted to specific requirements by its coherent and effective plug-in environment.

A simulation with *Pasimodo* usually consists of the main components, particles, interactions and integrators. The main simulation components used in this work are briefly discussed below. In addition, there are auxiliary objects like frames of reference, particle filters or data output.

4.1.2.1 Particles

The particles used to model granular materials are *spheres*, either with three (only translation) or six (translation and rotation) degrees of freedom (DOF). A sphere is defined by radius, mass and its state by position, velocity and angular velocity in the case of 6 DOF. Another important group of particles are *triangles* that may be used to define triangulated surfaces such as walls or structures. In general they are massless. To obtain particles of arbitrary shape, triangles may be aggregated as particle compounds which also carry mass. Spheres and triangles may also be termed DEM particles (see chapter 4.3).

The third kind of particles, so-called *SPH-particles*, are used to represent a fluid, e.g. water. This approach fundamentally differs from that for granular materials using spheres, discussed in detail in chapter 4.2.

4.1.2.2 Interactions

In dynamical systems as considered in this work, the interaction between the different kinds of particles is decisive for the behaviour of the systems and thus the reproduction of the physical processes. The constitutive nature of the interactions controls the overall behaviour of the modelled materials. Thus, the kind of interaction and its parameterisation may be crucial for the outcome of a simulation.

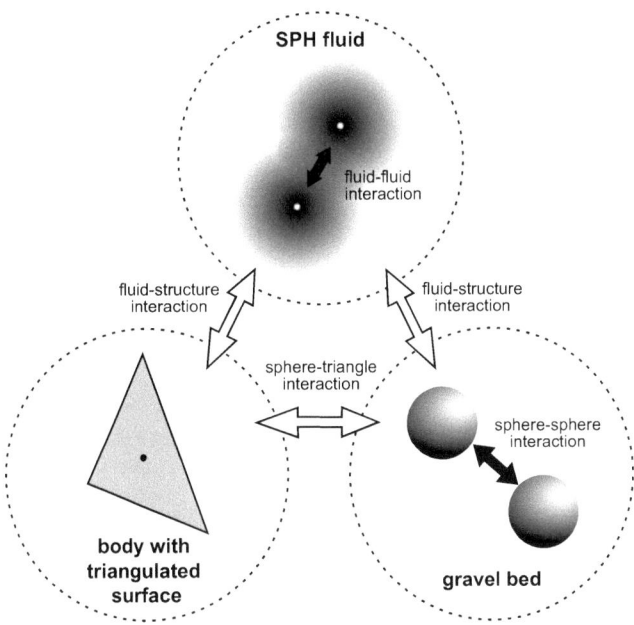

Fig. 4-2: Different kinds of particles and interactions applied in this work.

The different interactions applied in this work are depicted in Fig. 4-2. They may be categorised into 'internal' and 'external' interactions corresponding to the matter they actually represent and the interaction with other materials. 'Internal' refers to the interactions between SPH particles inside the fluid, i.e. fluid-fluid interaction, or between spheres with 6 DOF that represent the gravel bed (sphere-sphere interaction). The SPH interaction consists actually of the discretisation of the governing fluid equations by the use of a smoothing kernel and its influence on neighbouring particles (see chapter 4.2). For the interaction between spheres different force laws can be applied to prevent penetration; they are described in section 4.3.2 in detail. 'External' interactions on the one hand are sphere-triangle interactions that are treated in a similar way as sphere-sphere interactions, but the triangular particle is usually regarded as fixed, because triangles are used to model walls or static structures in this work. On the other hand, fluid-structure interactions are termed external comprising the interaction between SPH fluid particles and spheres or triangles. Details can be found in section 4.4.

4.1.2.3 Integrators

Another basic component of dynamic simulations is the integrator. It is responsible for the advancement of the particles in time by discrete time steps based on various stability criteria. For the time integration of the motion of DEM particles, several integrators are available; their application depends on the particularity of the problem and on accuracy requirements. For hybrid simulations with SPH and DEM particles as applied in the current work, for both methods the same kind of integrator is preferred to prevent problems due to asynchronism. For details see section 4.2.7 and 4.3.5.

4.2 Smoothed Particle Hydrodynamics

4.2.1 Introduction

4.2.1.1 General

Monaghan (2005b) describes the basic idea behind SPH as replacing the fluid by a set of points that follow the motion of the fluid and carry information about the properties of the fluid. These points can be seen either as interpolation points for the discretisation of the governing equations or as real material particles. This approach for fluid flow, where particles typically have fixed mass, allows for the advection of contact discontinuities while preserving Galilean invariance and reducing computational diffusion of various fluid properties including momentum.

Monaghan (1994) applied the method to free surface flows and demonstrated that SPH requires no explicit treatment of the free surface. In contrast, other methods like finite difference or finite volume schemes need special approaches that would require very fine meshes or adaptive grids for the modelling of complex flow with one or several convoluted free surfaces. Furthermore, the interaction with rigid bodies or boundaries can be handled as particle to particle interaction without the need of additional tracking or capturing of the movable interface. Overviews about SPH can be found in Monaghan (2005a), Monaghan (1992) or Liu and Liu (2003) for example.

Compared to established numerical schemes like FDM, the SPH method is still under development. It has been improved by contributions of many researchers during the last two decades and the number of applications increases continuously. Nevertheless, one of the main and well-recognised drawbacks is the high computational cost when it comes to 3D applications, especially when a fine special resolution is desired (Gomez-Gesteira et al. (2010b)).

4.2.1.2 Different Kinds of SPH

The standard SPH method (Monaghan (2005a)) used in this work is also termed "*weakly compressible SPH*" (WCSPH), because the computation of the pressure is based on an equation of state for water. This approach is suitable for flows where relative density variations range within 1% (see Monaghan (1992)). Some draw backs are for instance pressure fluctuations and very long computation times. The latter are due to the used CFL condition that depends on the velocity of sound instead of the fluid velocity. The velocity of sound is usually many times larger than the maximum velocity of the fluid.

4.2 Smoothed Particle Hydrodynamics

To circumvent these problems, a different approach has been introduced by Cummins and Rudman (1999) for flows without free surfaces and has been extended by Shao and Lo (2003) to free surface flows. This alternative approach is often termed *"truly incompressible SPH"* (ISPH). Instead of an equation of state, a Poisson equation is used to predict the pressure. For the discretisation of the Laplacian in the pressure Poisson equation a combination of a standard SPH derivative and a finite difference approximation may be applied. The approach to solve the Poisson equation is similar to grid-based Navier-Stokes solvers, i.e. application of the Chorin-type projection method (Chorin (1968)) and suitable linear solvers such as Bi-CGSTAB. Another method to cope with incompressibility without the need of solving a Poisson equation for the pressure is presented by Ellero *et al.* (2007). Their model requires that the volume of the fluid particles remains constant which is obtained by the solution of an additional set of non-linear equations.

The different approaches to treat the compressibility of the fluid are still an open topic in the SPH community. By comparison of the approaches, Hughes and Graham (2010) reach the conclusion that WCSPH performs as well as ISPH does and in some respects even better. Alternatively, Lee *et al.* (2010) show that ISPH is superior for some cases.

4.2.2 Representation of Fluids by Particles

4.2.2.1 Particle Approximation

Since SPH is a Lagrangian method for fluid dynamics, each particle i with position vector $\vec{r}_i$ moves with the fluid flow and carries quantities such as the velocity $\vec{u}_i$, the density $\rho_{f,i}$ and its mass m_i. In other words, these quantities are only known at the location of the particle itself. From a mathematical point of view, this situation can be represented with the help of the mathematical construct of Dirac, the Dirac delta function[5], $\delta(x)$, defined as

$$\delta(x) = \begin{cases} \infty & \text{if } x = x_0 \\ 0 & \text{otherwise} \end{cases} \quad ; \quad \int \delta(x_0 - x)dx = 1 \ . \tag{4.1}$$

Actually, the Dirac delta is not a strict function, but can be approximated by a well behaved function that has the following properties:

$$\lim_{h \to 0} W(x,h) = \delta(x) \ , \tag{4.2}$$

$$\int W(x_0 - x, h)dx = 1 \ , \tag{4.3}$$

where W is called kernel function and integration is over $\mathbb{R}^\sigma$, where σ is the dimension. An example for such a kernel function in the form of a Gaussian is given in Fig. 4-3, where h is the smoothing length. For decreasing h, the function tends to a single peak while the area or volume below the function course remains equal to unity, which corresponds to conditions (4.2) and (4.3).

---

[5] The Dirac delta function has an infinitely large value at its origin x_0 where $x = 0$ and is zero everywhere else. A further property is that its total integral over the space of its definition is equal to 1.

4 Numerical Methods

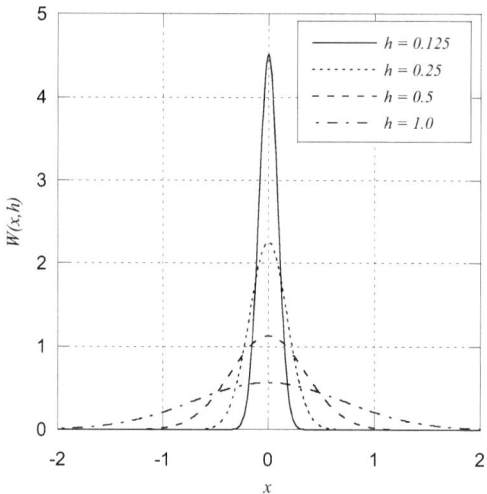

Fig. 4-3: Gaussian kernel function depending on smoothing length h, here in one dimension. The volume below each curve is equal to 1.

Based on equation (4.3) an interpolation at $\vec{r}_0$ for any quantity or function $A_r(\vec{r})$ can be obtained by the integral interpolant

$$A_I(\vec{r}_0) = \int A_r(\vec{r}) W(\vec{r}_0 - \vec{r}, h) d\vec{r} \ . \tag{4.4}$$

For numerical discretisation, the integral interpolant has to be approximated by a summation interpolant. Considering particles (interpolation points) with mass m, density ρ and position $\vec{r}$ identified by indices a and b, where a identifies the particle of interest and b the neighbouring particles with masses according to a volume element of the fluid $m_b = \rho_b(\vec{r}_b) d\vec{r}_b$ and $A_b = A_r(\vec{r}_b)$, the summation interpolant can be written as

$$A_a(\vec{r}_a) = \sum_b m_b \frac{A_b}{\rho_b} W(\vec{r}_a - \vec{r}_b, h) \ . \tag{4.5}$$

For example, the density can be estimated by

$$\rho_a(\vec{r}_a) = \sum_b m_b W(\vec{r}_a - \vec{r}_b, h) \ . \tag{4.6}$$

By using a kernel function that is differentiable, the derivative of A_a can be obtained by ordinary differentiation as

$$\vec{\nabla} A_a(\vec{r}_a) = \sum_b m_b \frac{A_b}{\rho_b} \vec{\nabla}_a W_{ab} \ . \tag{4.7}$$

For the sake of clarity, the notation $\vec{\nabla}_a W_{ab}$ was introduced to denote the gradient $\vec{\nabla} W(\vec{r}_a - \vec{r}_b, h)$ taken with respect to the position of particle a. Since the derivative in form of equation (4.7) is not very accurate even for a constant function, it should not be used for practical applications. According to Lehnart (2008), other forms of the derivative are used that are more accurate; they depend on the properties of the equation to be discretised. Their application to the Euler equations for fluid flow is introduced in section 4.2.3.1.

4.2.2.2 Kernel Functions

A general formulation of a kernel function is provided by Morris *et al.* (1997):

$$W(r_{ab}, h) = \frac{1}{h^\sigma} f\left(\frac{r_{ab}}{h}\right), \tag{4.8}$$

where σ is the dimension of the system, h is the smoothing length and $r_{ab} = |\vec{r}_a - \vec{r}_b|$ is the distance between particles a and b. Note that the dimension of the kernel function is 1/length$^\sigma$.

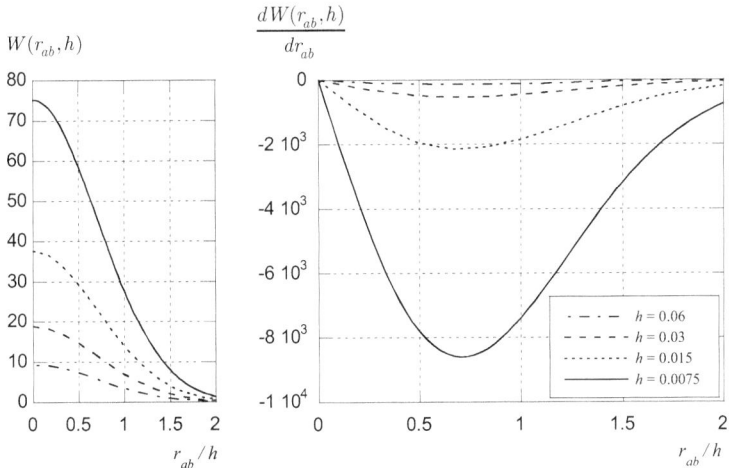

Fig. 4-4: Gaussian kernel function and its derivative for different smoothing lengths h.

According to Monaghan (1992), the first golden rule of SPH is the use of a Gaussian as kernel function when a physical interpretation of an SPH equation is desired. The Gaussian kernel is given by

$$W(r_{ab}, h) = \alpha_\sigma e^{-r_{ab}^2/h^2} \tag{4.9}$$

where α_σ is $1/(h\pi^{1/2})$, $1/(h^2\pi)$ and $1/(h^3\pi^{3/2})$ in one-, two- and three-dimensional space, respectively (index $\sigma \in \{1, 2, 3\}$ denotes the dimension of the problem). For the one-dimensional case, the derivative of the Gaussian reads

4 Numerical Methods

$$\frac{dW(r_{ab},h)}{dr_{ab}} = -\frac{2r_{ab}}{h^2} W(r_{ab},h) \ , \tag{4.10}$$

and its second derivative is

$$\frac{d^2W(r_{ab},h)}{dr_{ab}^2} = -\frac{2(h^2 - 2r_{ab}^2)}{h^4} W(r_{ab},h) \ , \tag{4.11}$$

which vanishes at $r_{ab} = h/\sqrt{2}$ and also holds for the 2D and 3D case.

Equations (4.9) and (4.10) are depicted in Fig. 4-4 for different smoothing lengths depending on the ratio r_{ab}/h. The Gaussian kernel has no compact support, i.e. it will not vanish outside a finite interval, but it rapidly falls off with distance and tends to zero for $r_{ab}/h \geq 2$. Other popular kernel functions are kernels based on splines (see e.g. Monaghan and Lattanzio (1986)) that have compact support. The Gaussian kernel has proved to be a good choice with regard to accuracy and efficiency and has been successfully applied in many simulations. For the present work, the Gaussian kernel with a cut-off at distance $r_{ab} = 2h$ is preferred.

4.2.3 Governing Equations

4.2.3.1 Euler Equations and Discretisation

By discretisation of the fluid with SPH particles, the equations describing fluid flow as a continuum, as the Euler equations in section 3.2.4.2, can be written in Lagrangian form according to section 3.2.4.3. Thus, the original partial differential equations reduce to a set of ordinary differential equations which can be discretised for particles according to the approaches introduced in section 4.2.2.1, i.e. index a denotes the actual particle and index b its neighbours within the cut-off distance. The properties of particle a are mass m_a, density ρ_a, velocity $\vec{u}_a$ and position $\vec{r}_a$ and similar for neighbouring particles with index b.

The conservation of mass reads

$$\frac{D\rho_f}{Dt} = -\rho_f \vec{\nabla} \cdot \vec{u} \ , \tag{4.12}$$

and can be discretised according to equation (4.7) as follows:

$$\hat{D} = \frac{D\rho_a}{Dt} = \rho_a \sum_b \frac{m_b}{\rho_b} (\vec{u}_a - \vec{u}_b) \cdot \vec{\nabla}_a W_{ab} \ . \tag{4.13}$$

According to Monaghan (2005a), this approach also allows for accurate results in the presence of strong density variations.

4.2 Smoothed Particle Hydrodynamics

Conservation of momentum

$$\frac{D\vec{u}}{Dt} = -\frac{1}{\rho_f}\vec{\nabla}p + \vec{f}_e \qquad (4.14)$$

leads to its discretised form

$$\hat{F} = \frac{D\vec{u}_a}{Dt} = -\sum_b m_a \left(\frac{p_a}{\rho_a^2} + \frac{p_b}{\rho_b^2}\right)\vec{\nabla}_a W_{ab} + \vec{f}_e . \qquad (4.15)$$

The particles are moved by

$$\frac{D\vec{r}_a}{Dt} = \vec{u}_a , \qquad (4.16)$$

and its discrete form will be introduced in the section on time integration, page 61.

Equations (4.13) and (4.15) are the Euler equations discretised by the SPH method according to Monaghan (1992). The equation system is closed by an appropriate equation of state for the pressure p (see section 4.2.3.2) and can be solved with initial conditions and advanced in time by the integration scheme presented in section 4.2.7.1.

4.2.3.2 Equation of State

Since the speed of sound in water is usually large compared with fluid velocities, i.e. the Mach number is quite low, the flow of water is modelled as incompressible (see section 3.2.4.1 on incompressible Navier-Stokes equations). However, for SPH, the motion of the fluid particles is simulated based on the compressible Euler equations, i.e. particles may be regarded as the molecules of a gas and their motion is driven by local density gradients. According to the laws of thermodynamics, the pressure can be related to the density by an equation of state for a compressible fluid to close the governing equations. Thus a quasi-incompressible equation of state is used for SPH. It may be noted that this approach of considering a compressible fluid is similar to Chorin's artificial compressibility method (Chorin (1967)) that can be used for the grid-based numerical solution of the incompressible Navier-Stokes equations.

For free surface flows, an equation of state similar to that introduced in section 3.2.4.3 is often applied. By omitting the addition of one in the dominator for large values of B, Equation (3.25) can be rewritten as

$$p_a = B\left[\left(\frac{\rho_a}{\rho_0}\right)^{\gamma_p} - 1\right], \qquad (4.17)$$

where ρ_0 is the reference density of the fluid, ρ_a is the particle density and usually $\gamma_p = 7$. Actually, the parameter γ_p (between 1 and 7) and the form of the equation of state may be chosen depending on the problem at hand. The use of equation (4.17) with large values of γ_p causes pressure to respond strongly to variations in density which may not be desired for low Reynolds number flows. Thus, for such situations $\gamma_p = 1$ and a modified form of the equation of state $p_a = c_s^2 \rho_a$ is preferred, as suggested e.g. by Morris et al. (1997).

In equation (4.17) the choice of B determines the speed of sound c_s. Since the time-step size of the simulation may depend on the speed of sound (as introduced later in section 4.2.7.2), a rather small value of c_s compared to its effective value of ~1500 m/s is preferred to gain a faster simulation progress. The parameter B can be related to c_s as follows

$$B = \frac{c_s^2 \rho_0}{\gamma_p} . \tag{4.18}$$

In order to limit density variations to a maximum of 1%, Monaghan (1994) argues that the sound velocity has been chosen so that the Mach number of the flow should be 0.1 or less; this yields

$$c_s = 10 u_{ref} , \tag{4.19}$$

i.e. $\mathrm{Ma} = u_{ref}/c_s = 0.1$. Nevertheless, Morris et al. (1997) obtained reasonable results for moderate flow conditions where density varies by at most 3%. It may be added that, although the compressibility effect for a Mach number of 0.1 is generally considered acceptable as a nearly-incompressible flow approximation, SPH behaves as a rarefied gas in regions of low pressure (see Issa et al. (2005)).

The reference velocity u_{ref} depends on the problem, i.e. for a dam break problem with initial water depth H_0 the approximate upper bound to the velocity is $u_{ref} = \sqrt{2gH_0}$ whereas for shallow water flows, where the ratio of wavelength to water depth tends to zero, the reference velocity is equal to the wave propagation velocity $u_{ref} = \sqrt{gh_f}$.

4.2.4 Enhancements

4.2.4.1 Artificial Viscosity

Since the fluid is considered to be inviscid in the present work (compare section 3.2.4.2), the term in the Navier-Stokes equations that depends on viscosity has been omitted. Nevertheless, the introduction of some sort of damping similar to finite difference schemes may be necessary for the stability of the numerical scheme. Therefore an artificial viscosity in terms of an artificial pressure is introduced in the momentum equation (see e.g. Monaghan (2005a)). This approach has been successfully applied for the simulation of shocks (e.g. Liu and Liu (2003) or Monaghan (2005b)) to avoid superficial oscillations of the velocity and pressure field.

The artificial viscosity term has two contributions. The first produces a bulk viscosity and is expressed as a volume-viscous pressure in relation to the velocity gradient:

$$q_v = \begin{cases} -\alpha \rho_f h c_s \vec{\nabla} \cdot \vec{u} & if \ \vec{\nabla} \cdot \vec{u} < 0, \\ 0 & otherwise, \end{cases} \tag{4.20}$$

4.2 Smoothed Particle Hydrodynamics

where h is the smoothing length. The second is based on the von Neumann-Richtmyer pressure and is necessary to handle high Mach number shocks where complete penetration of particles has to be reduced (see Monaghan (1985)),

$$q_v = \begin{cases} \beta \rho_f h^2 \left(\vec{\nabla} \cdot \vec{u}\right)^2 & \text{if } \vec{\nabla} \cdot \vec{u} < 0, \\ 0 & \text{otherwise}. \end{cases} \quad (4.21)$$

In the equations above $\alpha, \beta \geq 0$ are constants. Even though the values of α and β are not critical, good results were obtained for free surface flows by a choice of $\alpha = 0.01$ and $\beta = 0$ (compare Monaghan (1994)). The above expressions are introduced in the momentum equation by $\Pi_{ab} \sim q_v/\rho_f^2$ implying

$$\frac{D\vec{u}_a}{Dt} = -\sum_b m_b \left(\frac{p_a}{\rho_a^2} + \frac{p_b}{\rho_b^2} + \Pi_{ab} \right) \vec{\nabla}_a W_{ab} + \vec{f}_e \, . \quad (4.22)$$

According to Monaghan and Gingold (1983), the usual discretisation approaches for the velocity gradient as used for the discretisation of the Euler equations are leading to unsatisfying results with oscillations especially for shock tube problems. As alternative they suggest the use of

$$\vec{\nabla} \cdot \vec{u} = \frac{(\vec{u}_a - \vec{u}_b) \cdot (\vec{r}_a - \vec{r}_b)}{(\vec{r}_a - \vec{r}_b)^2 + \eta_v^2} \quad (4.23)$$

for the discretisation of equations (4.20) and (4.21), where η_v^2 prevents singularities and $\eta_v \ll h$. For the current work $\eta_v = 0.1h$ as suggested by Monaghan (2005a). To ensure symmetry of viscosity and conservation of momentum the sound velocity c_s and density ρ_f are determined by averaging,

$$\bar{c}_{ab} = \frac{1}{2}(c_{s,a} + c_{s,b}), \quad \bar{\rho}_{ab} = \frac{1}{2}(\rho_a + \rho_b) \, . \quad (4.24)$$

Combination of the two artificial pressure terms, i.e. equations (4.20) and (4.21), and application of expressions (4.23) and (4.24) leads to the discrete form of the artificial viscosity term,

$$\Pi_{ab} = \begin{cases} \dfrac{-\alpha \bar{c}_{ab} \mu_{ab} + \beta \mu_{ab}^2}{\bar{\rho}_{ab}} & \text{if } (\vec{u}_a - \vec{u}_b) \cdot (\vec{r}_a - \vec{r}_b) < 0, \\ 0 & \text{otherwise}, \end{cases} \quad \alpha, \beta \geq 0, \quad (4.25)$$

where

$$\mu_{ab} := \frac{h(\vec{u}_a - \vec{u}_b) \cdot (\vec{r}_a - \vec{r}_b)}{(\vec{r}_a - \vec{r}_b)^2 + \eta_v^2} \, . \quad (4.26)$$

4 Numerical Methods

For low Reynolds number flows or fluids with large viscosity differences real viscosity as appearing in the Navier-Stokes equations may be included according to Morris *et al.* (1997) (see e.g. Issa *et al.* (2005) or Monaghan (2006) for applications).

4.2.4.2 Artificial Stress

Fig. 4-5: Tensile instability observed at a backward facing step. Flow direction is from left to right and acuity indicates the occurrence of negative pressure.

The forces between SPH particles in a fluid are derived from the momentum equation depending on pressure and the first derivative of the kernel (see equation (4.15)). As depicted in Fig. 4-4, the kernel derivative has its maximum at a particle distance larger than zero. Thus, if particles come very close, the interaction force between them will decrease. This may lead to unphysical clumping of particles (see Fig. 4-5) and is called tensile instability.

According to Monaghan (2000), this instability can be removed by adding an artificial pressure term $R_{ab} f_{ab}^n$ to the momentum equation,

$$\frac{D\vec{u}_a}{Dt} = -\sum_b m_b \left(\frac{p_a}{\rho_a^2} + \frac{p_b}{\rho_b^2} + \Pi_{ab} + R_{ab} f_{ab}^n \right) \vec{\nabla}_a W_{ab} + \vec{f}_e . \tag{4.27}$$

The function f_{ab}^n increases as the particle distance decreases

$$f_{ab}^n = \frac{W(r_{ab}, h)}{W(\Delta s, h)} , \tag{4.28}$$

where W is the kernel function, r_{ab} the distance between particles a and b, h is the smoothing length and Δs the initial particle spacing. For fluid dynamical simulations $n = 4$. For negative pressure, the factor R_{ab} has contributions of both interacting particles

$$R_{ab} = R_a + R_b \tag{4.29}$$

and depends on the pressure in the following way:

$$R_i = \begin{cases} \varepsilon_s |p_i|/\rho_i^2 & \text{if } p_i < 0, \\ 0 & \text{otherwise}, \end{cases} \quad i = a,b, \tag{4.30}$$

where $\varepsilon_s = 0.2$ is a typical value. Artificial stress is also introduced for positive pressure to prevent the formation of local linear structures in liquids. Hence, for $p_a > 0$ and $p_b > 0$ the factor R_{ab} is parameterised as

$$R_{ab} = 0.01 \left(\frac{p_a}{\rho_a^2} + \frac{p_b}{\rho_b^2} \right). \tag{4.31}$$

In the example of a backward facing step depicted in Fig. 4-5, where a tensile instability was observed, the particle cluster stays intact after formation and moves along with the flow. A value of $\varepsilon_s > 0.5$ was necessary to prevent the occurrence of the instability in the given example.

4.2.4.3 Correction for Free Surface Flows

The intuition of Monaghan (1989) to prevent penetration of fluids impinging each other is termed XSPH (where "X" is the unknown factor). The XSPH correction is useful to obtain better results for free surface flows (see Monaghan (1994)) or for immiscible multiphase flows. A correction for the velocity is introduced that leads to an adaptation of the particle velocity to the mean velocity of the surrounding particles, which keeps the particles to move more orderly. The correction term, added to the right-hand side of equation (4.16) is

$$\Delta \vec{u}_a = \varepsilon_X \sum_b m_b \frac{(\vec{u}_b - \vec{u}_a)}{\overline{\rho}_{ab}} W_{ab}, \tag{4.32}$$

where $\overline{\rho}_{ab} = (\rho_a + \rho_b)/2$. The parameter $0 \leq \varepsilon_X \leq 1$ was introduced by Monaghan (1992) and is usually chosen as $\varepsilon_X = 0.5$.

4.2.4.4 Variable Smoothing Length

The smoothing length h controls the number of neighbouring particles that contribute to the solution of the momentum and continuity equations and thus exerts an influence on the accuracy of the simulation. Thus, the idea is to introduce a smoothing length that varies according to local conditions, and increased spatial resolution seems to be obvious (Hernquist and Katz (1989)). In areas with higher particle density a smaller smoothing length is necessary than in regions with small particle density; this leads to a definition of the particle smoothing length h_a depending on its density ρ_a

4 Numerical Methods

$$h_a \propto \left(\frac{1}{\rho_a}\right)^{1/\sigma} ; \qquad (4.33)$$

here σ is the number of spatial dimensions. According to Benz (1990), a simple approach for the time rate of change of the smoothing length is by using the time derivative of equation (4.33),

$$\frac{Dh_a}{Dt} = -\left(\frac{h_a}{\sigma \rho_a}\right)\frac{D\rho_a}{Dt} . \qquad (4.34)$$

Since the kernel function depends on the smoothing length, the time derivate of the kernel function has to be considered in the momentum and continuity equations, leading to an iterative solution procedure. A simpler and explicit approach is to use simple averaging of the kernels and kernel gradients (Monaghan (1992)

$$W_{ab} = \frac{1}{2}\left(W\left(r_a - r_b, h_a\right) + W\left(r_a - r_b, h_b\right)\right) , \qquad (4.35)$$

$$\vec{\nabla}_a W_{ab} = \frac{1}{2}\left(\vec{\nabla}_a W\left(r_a - r_b, h_a\right) + \vec{\nabla}_a W\left(r_a - r_b, h_b\right)\right) . \qquad (4.36)$$

Price (2004) used this simplified approach for some of his test problems and obtained satisfactory results with a substantially improved accuracy.

4.2.4.5 Turbulence Models

The standard formalism of SPH was successfully applied to complex flow types such as wave breaking, e.g. by Landrini *et al.* (2007). It was shown that detailed properties of vortices can be recovered. According to Cottet (1996), artificial viscosity models can be seen as eddy viscosity models but parameters have no explicit reference to any regularization of motion, i.e. the parameters have to be calibrated according to the problem at hand. However, in general the approach allows for taking into account turbulent effects in a similar way as algebraic turbulence models.

An advanced turbulence model is presented by Monaghan (2002). He introduced a fully Lagrangian turbulence model based on the Lagrangian averaged alpha model that is a powerful extension to the XSPH algorithm and is similar to the Large Eddy Simulation (LES) method.

Other advanced approaches are based on the adaptation of traditional turbulence models to the SPH method. The use of Reynolds-averaged flow equations is very common in engineering. This approach can also be adopted for SPH equations and has been successfully used recently for various applications, especially wave braking, with different eddy viscosity models such as the mixing length closure or the k-ε model (see e.g. Violeau and Issa (2007a), Shao (2006)). Also the concept of LES is applied to the SPH method in terms of the sub-particle scale (SPS) turbulence model introduced by Shao and Gotoh (2005). An overview of these approaches is given by Violeau and Issa (2007b).

4.2.5 Initial Conditions

Simulations with SPH are initialised by providing appropriate values for all particles. Thus, the following properties of a particle a are set at the beginning of the simulation: the mass m_a that is the only constant property of a particle (besides the kernel for constant h), the density of the particle $\rho_{a,0}$ and its initial velocity $\vec{u}_{a,0}$, where the index 0 denotes initial values. Furthermore, the initial particle spacing Δs and the initial time step are important values.

If the fluid is considered to have no initial pressure variation, the mass of a particle can be obtained by

$$m_a = \rho_0 (\Delta s)^\sigma , \qquad (4.37)$$

in which for water usually $\rho_0 = 1000$ kg/m³. For free surface flows, such as open channel flows, a hydrostatic pressure distribution is usually a good choice as initial condition. The hydrostatic pressure condition reads

$$p(z) = \rho_0 g (H_0 - z) , \qquad (4.38)$$

where H_0 is the initial water depth. By combination of (4.38) with the equation of state (4.17), i.e. setting $p(z) = p_a$, the initial density of particle a at height z corresponding to a hydrostatic pressure distribution can be set as (compare Lehnart (2008))

$$\rho_{a,0} = \rho_0 \left(1 + \frac{\rho_0 g (H_0 - z)}{B} \right)^{1/\gamma_p} . \qquad (4.39)$$

By evaluating equation (4.37) for the resulting density, the mass of the particles is obtained.

It may be noted that due to difficulties in obtaining appropriate open boundary conditions for SPH models (see next section), initial conditions are applied instead in many cases, e.g. a tank filled with water and a bottom outlet which generates an inflow to a subsequent model section.

4.2.6 Boundary Conditions

To set or obtain appropriate boundary conditions for particle methods is a demanding task. For grid based methods the values at the boundary can be obtained directly, either by explicit definition of the value of a variable at the boundary or by use of boundary cells. The latter approach can also be applied to particle methods in the sense of ghost particles as used at rigid boundaries for example (see section 4.2.6.3). For open boundaries, such as in- and outflows, neither approach is applicable because the particles themselves define the computational domain and its boundary, i.e. particles have to be created or erased at open boundaries.

4.2.6.1 Inflow Boundary

A simple inflow boundary condition is to continuously generate particles at the boundary with a given velocity as depicted in Fig. 4-6. Since particles carry density, this is actually a mass flow and not a velocity boundary condition.

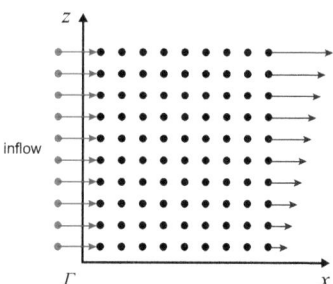

Fig. 4-6: Example of channel inflow as boundary condition: constant mass flow over boundary and velocity profile at distant section of the channel.

However, this approach is rather tricky, because regardless of the positions and momenta of the particles already existing in the domain new particles are created. This may lead to severe instabilities close to the boundary, if positions of existing and newly created particles become very close. Thus, it is recommended to choose the parameters and geometry of the inflow in a way that the flow close to the boundary is as continuous as possible (see e.g. section 5.4.2). As an alternative, applications with SPH often use a water tank with an outlet as inflow boundary.

4.2.6.2 Outflow and Pressure Boundary

There is no true outflow boundary for particle methods, except the removal of particles. Such a boundary consists of a defined domain where particles are erased as soon as they enter it, i.e. a particle sink. The extent of the domain has to be chosen sufficiently large according to the velocity of the particles close to the boundary and the size of the time step to prevent the skipping of particle sink.

For particle methods such as SPH it would be very difficult to prescribe the pressure of particles at a preferred location as a boundary condition and to obtain the desired effect. Nevertheless, appropriate pressure boundaries are very important in hydraulic engineering. Thus, approaches to model outflow boundaries as applied in laboratory experiments such as a weir or a permeable wall are also useful for SPH. However, implementation of these approaches requires calibration.

4.2.6.3 Rigid Boundaries

Monaghan and Kajtar (2009) note that *"SPH is a flexible robust method for simulating problems involving fluids interacting with rigid or elastic bodies. Despite its widespread application, the appropriate modelling of boundaries, fixed or moving is still not clear."*

One of the major problems with SPH concerning fluid-structure interaction is the encounter of fluid particle with a rigid boundary; the interpolant will be incomplete if no particles are present that identify the boundary. Therefore, different solutions that include the creation of virtual boundary particles are available to avoid boundary problems. Gomez-Gesteira *et al.* (2010a) distinguish three different kinds of boundary particles: ghost particles, repulsive particles and dynamic particles.

The first approach involves the creation of a ghost or mirror particle when a fluid particle comes into contact with the boundary as proposed by Libersky *et al.* (1993). The ghost particle has the same density and pressure as the fluid particle, but its velocity is opposite to introduce a repulsive effect. According to Morris *et al.* (1997) this procedure works well for straight channels, but introduces density errors for curved surfaces.

The use of repulsive particles was introduced by Monaghan (1994). The boundary particles are fixed and exert a repulsive force according to a Lennard-Jones potential for a given distance to the fluid particle. This approach has been refined by Monaghan and Kos (1999) and Monaghan *et al.* (2003). A further modification of this approach includes the adjustment of the repulsive force according to the magnitude of the local flow depth as presented by Rogers and Dalrymple (2008). However, for irregular free surfaces this approach may lead to numerical instabilities since the flow depth is not easy to predict.

For dynamic particles, the boundary particles are initially arranged as fluid particles, i.e. also inside an obstacle, but their position does not change with time or is externally influenced. The density and the pressure of the dynamic particles are evaluated by the continuity equation and the equation of state as for the fluid. This approach was applied by Morris *et al.* (1997) to ensure no-slip conditions for laminar wake flow. Potapov *et al.* (2001) used this approach for the combination of the SPH method with the DEM, where dynamic particles were placed inside the DEM particles without affecting the density of the DEM particle. For open channel flows, Issa (2005) used several rows of fixed dynamic particles to achieve appropriate boundary conditions. However the main problem of this kind is the evolution from the initial condition, where fluid particles move away from the wall causing a pressure decrease. This leads to a so-called pseudo-viscosity where small groups of particles remain stuck to the boundary. Furthermore, this approach does not necessarily prevent the penetration of the boundary by fluid particles.

According to Gomez-Gesteira *et al.* (2010a) the creation of realistic boundary conditions remains an open topic in SPH methods. The approaches for rigid or wall boundary conditions applied in this work are actually a combination of the first two concepts. A detailed description in terms of fluid-structure interaction is given in section 4.4.1.

4.2.7 Time Integration and Solution Algorithm

4.2.7.1 Integration Scheme

The fluid particles are advanced in time by the solution of the Lagrangian form of the Euler equations, i.e. the continuity (4.12) and the momentum equation (4.14). Since both equations are ordinary differential equations, theoretically any stable time-stepping scheme for ordinary differential equations can be used. However, for non-dissipative systems Monaghan (2005b)

4 Numerical Methods

recommends the use of a symplectic Verlet integrator of second order accuracy since it conserves angular momentum exactly and conserves energy better in comparison with a Runge-Kutta scheme.

For dissipative systems, Lehnart (2008) implemented a predictor-corrector method based on the leapfrog scheme (PC-leapfrog) as presented by Monaghan et al. (2003). Introducing the variables of particle a at the beginning of time step n as $\hat{D}_a^n, \rho_a^n, \hat{F}_a^n, \vec{u}_a^n$ and $\vec{r}^n$, the predictor step can be described as

$$\vec{u}_{a,p} = \frac{\vec{u}_a^n + \Delta t \hat{F}_a^n}{1 + \gamma_{pc} \Delta t}, \tag{4.40}$$

$$\vec{r} = \vec{r}_a^n + \Delta t \vec{u}_a^n + \frac{1}{2}\Delta t^2 \hat{F}_a^n, \tag{4.41}$$

$$\rho_{a,p} = \rho_a^n + \Delta t \hat{D}_a^n, \tag{4.42}$$

where γ_{pc} controls the damping of the velocity that is usually set to $\gamma_{pc} = 0$ and Δt is the size of the time-step n that is determined by the conditions given in the next section. It may be noted that the position vector $\vec{r}$ is not corrected, i.e. $\vec{r}_a^{n+1} = \vec{r}_p = \vec{r}$. Based on the predicted density $\rho_{a,p}$ and velocity $\vec{u}_{a,p}$, the values of $\hat{F}_{a,p}$ and $\hat{D}_{a,p}$ are predicted by evaluation of the corresponding continuity and momentum equations. Finally, the corrected values for the new time-step can be obtained by

$$\vec{u}_a^{n+1} = \vec{u}_{a,p} + \frac{1}{2}\Delta t \left(\hat{F}_{a,p} - \hat{F}_a^n \right), \tag{4.43}$$

$$\rho_a^{n+1} = \rho_{a,p} + \frac{1}{2}\Delta t \left(\hat{D}_{a,p} - \hat{D}_a^n \right). \tag{4.44}$$

Now, the particles can be advanced to the new positions $\vec{r}_a^{n+1}$ and initialized with the new values for $\vec{u}_a^{n+1}$ and ρ_a^{n+1}.

4.2.7.2 Time-step size

For the SPH method applied here, three characteristic time scales exist, namely

$$\frac{h}{u_{ref}}, \quad \frac{h^2}{\nu}, \quad \sqrt{\frac{h}{a_{max}}} \tag{4.45}$$

The first corresponds to the general stability condition for numerical problems where advection is dominant, i.e. the CFL-condition (Courant et al. (1928)). It means that in the time step Δt a quantity must not advance further than a given length scale. For SPH, the relevant length scale is the smoothing length h and the reference velocity u_{ref} is the higher of the maximum flow velocity or the specified sound velocity c_s, i.e. $u_{ref} = \max\left(\left| \vec{u} \right|_{max}, c_s \right)$. The second and the third characteristic time scales restrict the time step to the maximum of acting internal and external forces, i.e. the viscous forces and the applied forces in terms of the maximum particle acceleration a_{max}, whereas the former is only relevant for flows with low Reynolds numbers.

According to these considerations, the size of a time step can be obtained by the assignment

$$\Delta t = \alpha_s \min\left(\frac{h}{u_{ref}}, \frac{h^2}{\nu}, \sqrt{\frac{h}{a_{\max}}}\right), \qquad (4.46)$$

where α_s is a safety factor similar to the CFL number. According to the results reported by other researchers, α_s lies in the range of 0.125 and 0.5 (see e.g. Lee *et al.* (2008) or Monaghan *et al.* (2003)). Values around 0.25 may be preferred for flows with strongly varying boundary forces like those in the present work.

4.2.7.3 Solution Algorithm

The solution algorithm for the SPH method can be outlined as follows:

I. set initial conditions
II. while time $t \leq t_{end}$
  1. set boundary conditions
  2. determine time-step size Δt according to (4.46)
  3. for each particle i:
     3.1. determine interacting particles j and compute right-hand sides of the continuity (4.13) and momentum equation (4.15), i.e. $\hat{F}$ and $\hat{D}$.
     3.2. compute new values for time $t + \Delta t$ according to the time integration scheme, i.e. equations (4.40) to (4.44)
  4. advance particles to their new positions and increment time by Δt

4.2.8 Considerations of Accuracy

According to Monaghan (2005b), the integral interpolant is approximated very accurately by the summation interpolant using a Gaussian kernel for equi-spaced particles. Furthermore, a smoothing length larger than the particle spacing, i.e. $h > \Delta s$, results in a very good approximation, but for $h < \Delta s$ the accuracy is poor. For 2D simulations Graham and Hughes (2008) show that with a ratio of $h/\Delta s = 1.5$ the approximations are convergent for regularly spaced particles and are less sensitive to particle disorder than with ratios of 1.25 or 1. Thus, the common choice of a Gaussian kernel and a smoothing length $h = 1.5\Delta s$ based on the initial particle spacing seems to be reasonable.

When the particles are disordered, as it will occur in almost any dynamic simulation with SPH, the errors could be estimated by a Monte-Carlo estimate. However, since the moderate disorder of particles occurring in SPH simulations is not comparable with the large fluctuations included in the Monte-Carlo estimate, the error is actually much smaller. Furthermore, due to the disorder depending on the dynamics, traditional error estimates as used for FDMs or FEMs are not applicable. Thus, the best way to determine the accuracy of a SPH simulation is by comparing the results with known solutions, experimental data or the outcome of other numerical schemes.

4 Numerical Methods

As far as consistency is concerned, an increasing number of particles N will not necessarily lead to a higher accuracy. According to Rasio (2000), also the number of neighbouring particles N_N has to be increased in a way that N increases faster than N_N while the smoothing length h decreases. Furthermore, he states that the SPH method is consistent in the limit when $N \to \infty$, $N_N \to \infty$ and $h \to 0$ and says that convergence can be accelerated significantly by increasing the smoothness of the kernel. This implies that particles must overlap at all times in order to guarantee the convergence of the method.

4.3 Discrete Element Method

4.3.1 Basic Concepts

4.3.1.1 Introduction

For simulations of interacting rigid bodies, the focus is on their contact and the balancing of the occurring contact forces. Cundall and Hart (1992) distinguish between hard contacts where interpenetration of the bodies is regarded as non-physical and soft contacts that allow for interpenetration (see Fig. 4-7). For solids, the first seems to be reasonable from a physical point of view since a collision results in surface deformation. However, a simulation model for hard contacts at least has to exactly track the moment of contact (the deformation of the surface would be a further task, if required). This usually requires the application of an iterative scheme. Hence, corresponding applications are commonly restricted to a rather small number of interacting bodies.

If soft contacts are considered, the interpenetration is regarded to be an equivalent for the surface deformation. The contact forces are related to the displacement or the amount of interpenetration δ in general. A well-known example for that is the Hertz contact theory (see 4.3.2.2) which describes the contact between two deformable spheres. Furthermore, the approach of soft contacts is the basic concept of the discrete element method, since it allows for stable and accurate interaction modelling of rigid bodies and can be applied to an almost unlimited number of particles as far as computational resources are available (see section 2.2.2 for an overview of current applications).

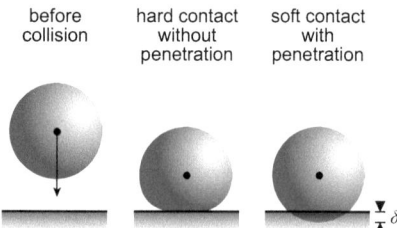

Fig. 4-7: Modelling approaches of interacting rigid bodies: collisions with and without penetration of the colliding bodies.

The procedure of a DEM simulation may be outlined as follows. In a first step it has to be detected for each particle whether collisions with neighbouring particles will take place or not, i.e. whether interpenetration occurs or not. If a collision occurs, a so-called penalty force depending on the

4.3 Discrete Element Method

amount of interpenetration will be applied. With regard to a pair of colliding particles, the penetration continues until the forces exerted by the particles are balanced by the penalty force, i.e. when maximum penetration is reached. The detailed approach includes different penalty force laws and is presented and illustrated with examples in the following sections.

4.3.1.2 Spring-Damper System

A common approach to model penalty forces between two colliding rigid objects is the implementation of a spring-damper system (e.g. Cundall and Strack (1979), Tsuji et al. (1992) and Tavarez and Plesha (2007)). For the case of two spheres, P_i and P_j, the state of the spring-damper system is defined depending on their overlapping or penetration depth δ, respectively

$$\delta = \left(r_i + r_j\right) - r_{ij} \quad \begin{cases} \leq 0 : inactive, \\ > 0 : active, \end{cases} \tag{4.47}$$

where r_i, r_j are the radii and r_{ij} is the distance between the two spheres. Please see appendix A.2 for general definitions used in this chapter. Fig. 4-8 shows an active spring-damper system.

The spring is responsible for putting back the spheres to the state of contact. It exerts a penalty force $\vec{F}_n\left(k(\delta)\right)$ depending on material properties and penetration depth δ in the direction of the spring-damper system axis $\vec{e}_{sd}$, i.e. normal to the contact surface,

$$\vec{F}_n\left(k(\delta)\right) = -k(\delta)\vec{e}_{sd} \,. \tag{4.48}$$

The penalty force $\vec{F}_n\left(k(\delta)\right)$ can be determined using different approaches, either linear or nonlinear depending on $k(\delta)$. Chapter 4.3.2 gives a brief overview.

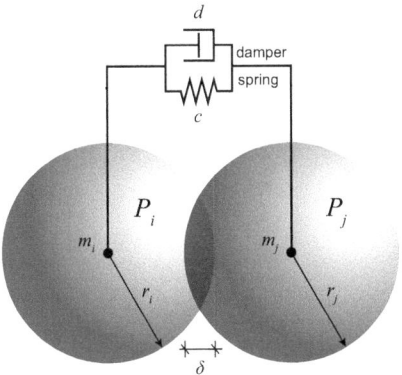

Fig. 4-8: Spring-damper system for the modelling of penalty forces due to overlapping.

A simple approach for modelling dissipation is the application of a viscous damper. The exerted force of the damper depends on the collision velocity $\dot{\delta} = \left|\vec{v}_i - \vec{v}_j\right|$ in the direction of the spring-damper system axis,

$$\vec{F}_d = -d\dot{\delta}\vec{e}_{sd} ,\qquad(4.49)$$

where d is the viscous damping coefficient with units [Ns/m].
By adding equations (4.48) and (4.49) the collision force $\vec{F}_c$ results as

$$\vec{F}_c = \vec{F}_n\left(k(\delta)\right) + \vec{F}_d .\qquad(4.50)$$

Assume that the spring-damper in Fig. 4-8 is fixed to the two horizontally aligned spheres with initial position $\delta = 0$. Physically, this is a one-dimensional problem. Furthermore, for a system of two bodies approaching each other, the reduced or effective mass is defined as (e.g. Brilliantov and Pöschel (2004))

$$\mu_m = \frac{m_i m_j}{m_i + m_j} .\qquad(4.51)$$

Applying Newton's second law to the colliding pair one obtains

$$\mu_m \ddot{\delta}\vec{e}_{sd} = \vec{F}_c .\qquad(4.52)$$

This leads to the following equation of motion depending on a penalty force model based on $k(\delta)$:

$$\mu_m \ddot{\delta} + d\dot{\delta} + k(\delta) = 0 \qquad(4.53)$$

4.3.2 Penalty Force Models

4.3.2.1 Linear Force Model

The most common model for determination of the force necessary to displace a spring is Hooke's law. The spring is assumed to be perfectly elastic and thus the model is considered to be conservative. For a spring with stiffness c [N/m] and displacement δ [m] acting in the direction of the spring-damper system axis $\vec{e}_{sd}$ the linear force law reads

$$\vec{F}_s(\delta) = -c\delta\vec{e}_{ij} = \vec{F}_n(k(\delta)) , \text{ i.e. } k(\delta) = c\delta .\qquad(4.54)$$

The behaviour of a corresponding spring-damper system can be illustrated by the one-dimensional problem depicted in Fig. 4-14 a), where the lower sphere P_j is fixed and the upper one has initial position $\delta = 0$ as well as mass $\mu_m = m_i$. Due to gravity $\vec{g} = (0,0,-g)^T$, the upper sphere S_i will oscillate around an equilibrium position for which

$$\vec{F}_g = \mu_m \vec{g} = -c\delta_0 \vec{e}_{ij} = \vec{F}_s(\delta) ,\qquad(4.55)$$

$$\delta_0 = \frac{\mu_m g}{c} \qquad(4.56)$$

holds. This leads to an equation equivalent to (4.53) but with displacement of the equilibrium position to δ_0

$$\zeta = \delta - \delta_0 , \tag{4.57}$$
$$\mu_m \ddot{\zeta} + d\dot{\zeta} + c\zeta = 0 . \tag{4.58}$$

Introducing the definitions

$$\gamma_d := \frac{d}{2\sqrt{\mu_m c}} \quad \text{and} \quad \omega_0^2 := \frac{c}{\mu_m} \tag{4.59}$$

one obtains

$$\ddot{\zeta} + 2\gamma_d \omega_0 \dot{\zeta} + \omega_0^2 \zeta = 0 \tag{4.60}$$

Equation (4.60) is a homogeneous linear second-order differential equation with constant coefficients that corresponds to a damped harmonic oscillator. Its solution can be characterized by the attenuation factor or damping ratio γ_d:

$0 < \gamma_d < 1$ weakly damped oscillations with angular frequency ω_d,
$\gamma_d > 1$ strong damping, no oscillations at all,
$\gamma_d = 1$ critical damping.

For no damping, i.e. $\gamma_d = 0$, the angular frequency of the system is ω_0 and the period T_o is obtained by

$$T_o = \frac{2\pi}{\omega_0} = 2\pi\sqrt{\frac{\mu_m}{c}} . \tag{4.61}$$

The ordinary frequency of the system is

$$f_0 = \frac{1}{T_o} = \frac{1}{2\pi}\sqrt{\frac{c}{\mu_m}} . \tag{4.62}$$

For the above mentioned cases, the response of a spring-damper system with a sphere S_i ($r = 0.05$ m, $\rho = 2800$ kg/m^3, $\mu \approx 1.5$ kg), spring constant $c = 36\mu\pi^2 \approx 521$ N/m and frequency $f_0 = 3$ is depicted in Fig. 4-9. The damping ratios are given as $\gamma_d = 0.025$ ($d \approx 1.4$ Ns/m) for weak damping $\omega_d \approx \omega_0$; $\gamma_d = 5$ ($d \approx 276.4$ Ns/m) for strong damping and $\gamma_d = 1$ ($d \approx 55.3$ Ns/m) for critical damping.

4 Numerical Methods

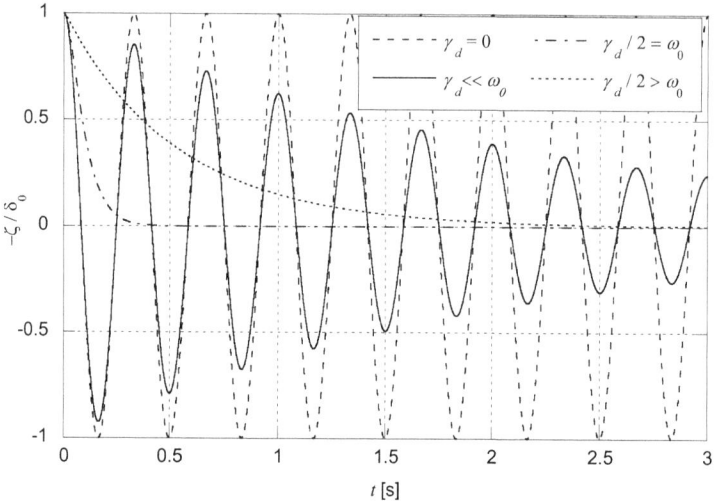

Fig. 4-9: Response of spring-damper system for various damping ratios.

A further discussion of the solution of equation (4.60) and examples can e.g. be found in Knudsen and Hjorth (2000).

4.3.2.2 Hertz Force Model

A more physically motivated approach for modelling the interaction of two perfectly elastic spheres with frictionless surfaces is based on the contact theory of Hertz (1882). The nonlinear force law is written as

$$\vec{F}_n(k(\delta)) = -K\delta^n \vec{e}_{ij} \, , \tag{4.63}$$

i.e. $k(\delta) = K\delta^n$, where K is the generalised stiffness constant. The exponent n depends on the distribution of the contact stresses and is set to 1.5, as in the original work by Hertz. For two colliding spheres i and j, the stiffness parameter depends on the radii and the material properties,

$$K = \frac{4}{3(\sigma_i + \sigma_j)} \left[\frac{r_i r_j}{r_i + r_j} \right]^{\frac{1}{2}} , \tag{4.64}$$

with material parameters σ_i and σ_j:

$$\sigma_k = \frac{1 - \nu_k^2}{E_k}, \quad (k = i, j) \, , \tag{4.65}$$

4.3 Discrete Element Method

where ν_k is Poisson's ratio and E_k is Young's modulus. An in-depth description of the Hertz contact theory is e.g. given by Popov (2010).

To compare the different approaches, the progression of the penalty force during impact for linear and Hertz law is depicted in Fig. 4-10.

The equivalent linear spring stiffness for a Hertzian contact can be obtained by its force potential. By integration of the penalty force (4.63) along the path of penetration ds one obtains the potential energy for the Hertz force law with $n = 1.5$ as

$$U_h = -\int_0^\delta \left|\vec{F}(k(\delta))\right| ds = \frac{2}{5} K \delta^{5/2} . \tag{4.66}$$

A similar equation holds for the force potential U_s of a displaced linear spring (see equation (4.80)). For identical maximum penetration the potentials are equal; this leads to

$$U_h = U_s = \frac{2}{5} K \delta_{\max}^{5/2} = \frac{1}{2} c_1 \delta_{\max}^2 . \tag{4.67}$$

Rearranging equation (4.67) one obtains the spring stiffness c_1 that is equivalent to the Herzian contact for similar maximum penetration $\delta_{\max}$

$$c_1 = \frac{4}{5} K \sqrt{\delta_{\max}} . \tag{4.68}$$

This implies that the application of the linear law with equivalent spring stiffness according to (4.68) instead of the Hertz law results in a 20% reduced penalty force at maximum penetration. This fact is also depicted in Fig. 4-10 by the curve denoted as "linear*", where the linear force is normalized by the maximum force corresponding to the Hertz solution. Furthermore, the impact time is smaller and because of the nonlinearity of the Hertz force law, a corresponding spring-damper system will not oscillate around its equilibrium position. This is apparent from Fig. 4-11, where in addition the penetration depth of the Hertzian contact, normalized by the equilibrium penetration depth of the linear spring $\delta_{0,lin}$, is shown; it implies similar penetration depths for both systems.

4 Numerical Methods

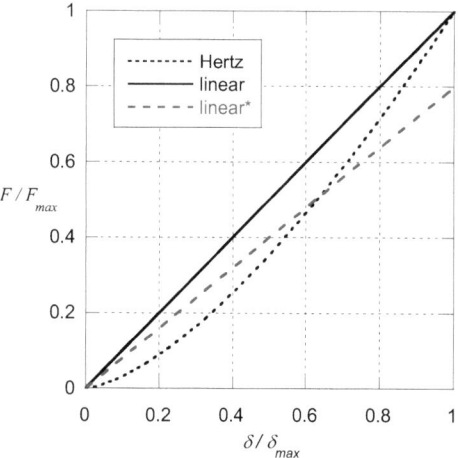

Fig. 4-10: Comparison of the linear and Hertz force laws, where $F = \left|\vec{F}_n(k(\delta))\right|$ and $F_{\max} = \left|\vec{F}_n(k(\delta_{\max}))\right|$, respectively. For the curve "linear*" the linear force is normalized by the maximum force corresponding to the Hertz solution.

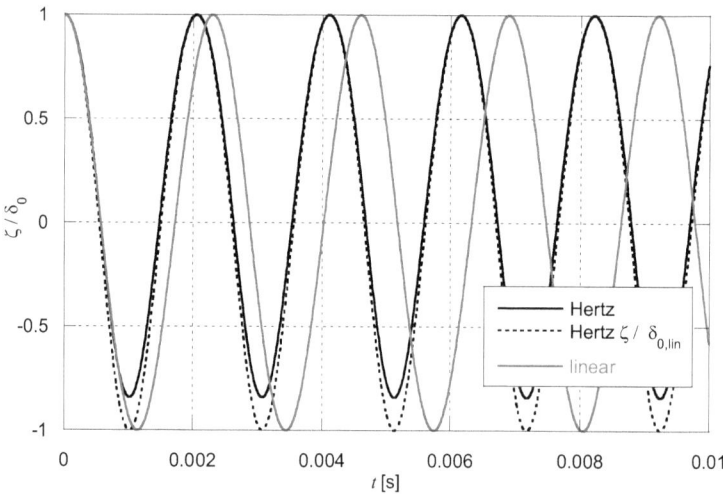

Fig. 4-11: Oscillation of a spring-damper with a linear and a Hertz force law and similar penetration depth

4.3 Discrete Element Method

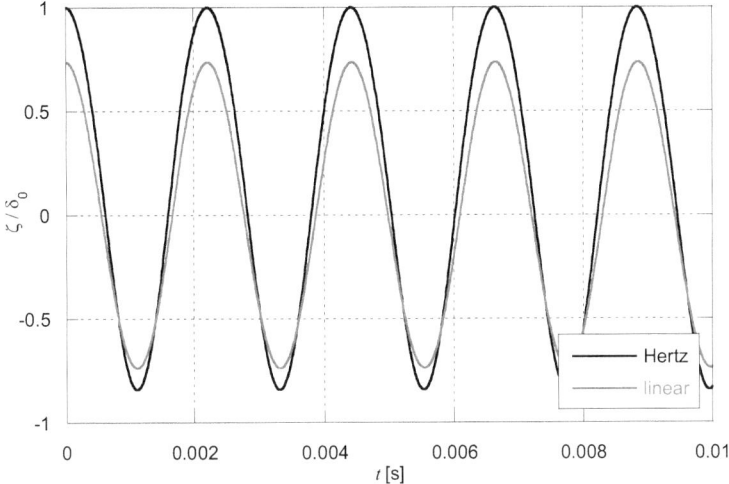

Fig. 4-12: Oscillation of a spring-damper with a linear and a Hertz force law and approximately similar frequency

Replacing the spring-damper system based on the Hertz force law by a harmonic oscillator with appropriate stiffness, a system with approximately the same frequency can be obtained. The momentum of an impact with duration t_c has to be equal for the linear and Hertz laws and is here estimated by a linear approximation depending on δ_{max} of the Hertz interaction

$$\int_0^{t_c} c_2 \delta dt = \int_0^{t_c} K \delta^{3/2} dt \approx \frac{t_c}{2} c_2 \delta_{max} = \frac{t_c}{2} K \delta_{max}^{3/2}, \quad (4.69)$$

implying

$$c_2 = K \delta_{max}^{1/2}, \quad (4.70)$$

where c_2 is the stiffness of a harmonic oscillator that has approximately the same frequency as a spring-damper with Hertz interaction given by the properties K and δ_{max}. An example of the approximation is depicted in Fig. 4-12. The approximated frequency of the linear law compared to the Hertz law is about 1% smaller for impact without velocity and about 5 to 10% larger for the more general case with impact velocity. The deviation slightly increases with smaller impact time.

4.3.2.3 Models with Dissipation

The penalty force models described in the previous sections do not consider energy dissipation during the process of impact. Observed energy loss is assumed to be due to material damping of the bodies, which would dissipate energy in the form of heat. Lankarani and Nikravesh (1994) studied impact in multibody systems and developed a contact model with hysteresis damping. They suggested separating the normal contact force into elastic and dissipative components. That approach corresponds to equation (4.53) with a nonlinear damping coefficient

$$d = \chi \delta^n \tag{4.71}$$

where n is set according to the contact law (see equation (4.63)) and the hysteresis factor is given by

$$\chi = \frac{3K(1-e_r^2)}{4v_c}, \tag{4.72}$$

in which v_c is the velocity at initial contact. The restitution factor e_r depends on the velocity before, v_c and after, v_c' contact and is defined as

$$e_r = -\frac{v_c'}{v_c}, \tag{4.73}$$

where $e_r = 1$ corresponds to an elastic and $e_r = 0$ to a completely inelastic collision. Further details on the restitution factor as well as a general formulation of a nonlinear spring-damper model can be found in Luding (1998).

4.3.2.4 Lennard-Jones Potential

Another approach to model the interaction of particles employs the Lennard-Jones Potential, that emerges from molecular dynamics (see e.g. Young (2002)). Compared to the former contact force models, not only a repulsive force is considered but also an attractive force is taken into account. The exerted force depends on the distance between the interacting particles, i.e. the force is not limited to contact in general. Due to its behaviour, the Lennard-Jones potential may be rather termed an interaction model than a contact force model. Furthermore, the Lennard-Jones potential is conservative.

For two interacting particles the Lennard-Jones potential is given as a function of their distance r_{ij} by

$$U(r_{ij}) = \alpha \varepsilon_p \left[\left(\frac{\sigma_p}{r_{ij}} \right)^n - \left(\frac{\sigma_p}{r_{ij}} \right)^m \right], \quad n > m, \tag{4.74}$$

where

$$\alpha = \frac{1}{n-m} \left(\frac{n^n}{m^m} \right)^{\frac{1}{n-m}},$$

(see e.g. Griebel et al. (2007)). The parameters of the potential are ε_p and σ_p. The depth of the potential, i.e. the intensity of the repulsive and attractive forces, can be given by ε_p [Nm], which may be also considered as the strength of a matter or material compound (see Fig. 4-13). For $\varepsilon_p = 0$ Nm or $\sigma_p = 0$ m the potential vanishes and $r_{ij} = 0$ m is physically not possible. Examples of two different potentials with $\sigma_p = 1$ m and depth $\varepsilon_p = 0.5$ Nm as well as $\varepsilon_p = 0.01$ Nm are depicted in Fig. 4-13.

4.3 Discrete Element Method

A common approach is to set $m = 6$ as used in the definition of the van der Waals force; moreover $n = 12$, however this choice does not stem from physical considerations but merely from considerations of mathematical simplicity, yielding $\alpha = 4$.

The force acting between two particles at distance r_{ij} can be obtained by differentiating equation (4.74). For $m = 6$ and $n = 12$ the derivative reads

$$\vec{F}_n(r_{ij}) = -\vec{\nabla} U(r_{ij}) = 24\varepsilon_p \frac{1}{r_{ij}}\left(\frac{\sigma_p}{r_{ij}}\right)^6 \left[1 - 2\left(\frac{\sigma_p}{r_{ij}}\right)^6\right] \vec{e}_{ij} \ . \tag{4.75}$$

The point where the force changes from repulsion to attraction, i.e. $\left|\vec{F}\right| = 0$, is given for non-zero values of ε_p, σ_p and r_{ij} as

$$r_0 = 2^{1/6}\sigma_p \ . \tag{4.76}$$

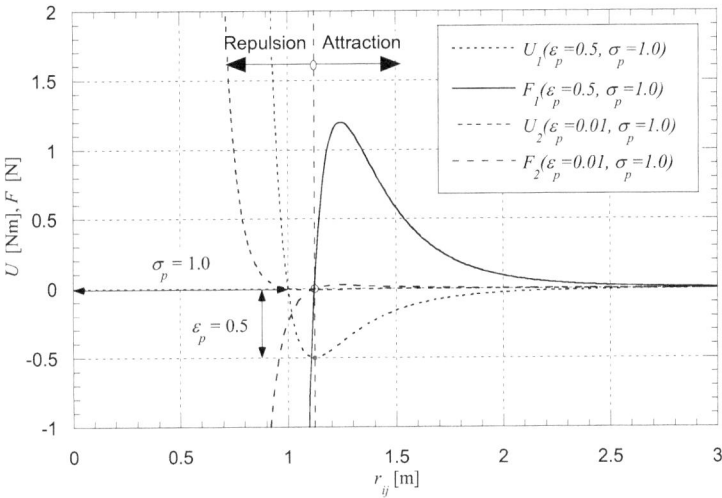

Fig. 4-13: Lennard-Jones potential for two different sets of parameters (potential: red, force: black)

Thus, equation (4.74) or (4.75) can be expressed in terms of r_0 by inserting $\sigma_p = 2^{-1/6} r_0$ (see appendix A.4.1). This is useful to provide appropriate initial conditions which correspond to the equilibrium state.

For non-molecular models with rigid hulls as considered in this work, the binary mode described by equation (4.47) is introduced to describe collisions. Furthermore, for a collision model a

4 Numerical Methods

formulation of the potential or force depending on the equilibrium penetration depth δ_{zero} is preferred.

$$r_0' = (r_i + r_j) - \delta_{zero} \,. \tag{4.77}$$

If $\delta_{zero} = 0$, only repulsive forces will act; this corresponds to a pure penalty force model; for $\delta_{zero} > 0$ also attraction may occur.

Due to the possibility of modelling attraction by use of the Lennard-Jones potential as interaction law, also adhesive forces of materials may be considered. However, for dry granular material they are not of importance, except e.g. for clays.

4.3.3 Modelling of Collisions

To illustrate the collision process, two examples how to calculate the penetration depth and thus the collision penalty force based on conservation of energy are given below. The cases discussed are depicted in Fig. 4-14. They represent the occurring kinds of collisions of the DEM model used in this work. On the one hand, the collision of a sphere with a fixed sphere is discussed (Fig. 4-14a), case A); this represents e.g. the interaction of a saltating sediment particle with a bed particle that is at rest. On the other hand, two colliding spheres with opposite velocity according to Fig. 4-14b) (case B) are studied. This case shows how the interaction of a sphere with a triangulated surface of a structure, e.g. a wall, is modelled and may be regarded as the general case of the interaction of two spheres where the exerted collision force reaches a maximum.

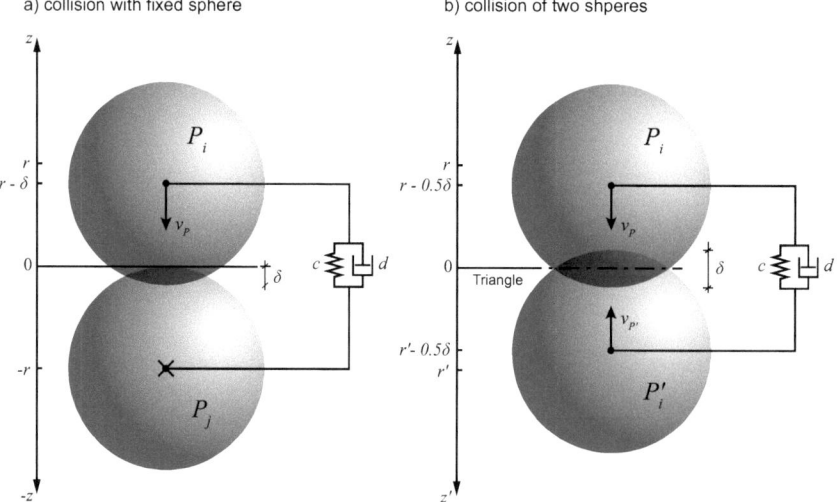

Fig. 4-14: Two different approaches for modelling the collision of a sphere with another rigid object, e.g. a wall or a sphere.

4.3 Discrete Element Method

Consider two spheres P_i and P_j with identical radii r, mass m and one degree of freedom (1 DOF) in the z-direction. Gravity is given as $\vec{g} = (0,0,-g)^T$. Displacement of the spring-damper system only takes place for positions $z_P(t) < r$. Furthermore, the systems are assumed to be conservative, i.e. dissipative forces such as damping and friction are neglected.

Case A: collision with fixed sphere

In this classic example, the sphere is released at a given height and after a certain time it hits the floor or wall, respectively. The sphere has an initial centre position $z_P(t_0) = z_0 = 1 + r$ and initial velocity $v_P(t_0) = 0$. At the starting time t_0 right before dropping the sphere only potential energy $U_0 = E_0 = mgz_0$ exists due to zero velocity. After a short free fall the sphere encounters the other rigid object at centre position $z_P(t_1) = z_1 = r$ and at time t_1. The initial impact velocity $v_P(t_1) = v_c$ at the time of contact of the sphere and the rigid object can be determined by evaluation of the system energy

$$E_1 - E_0 = U_1 + T_1 - E_0 = mgr + \frac{1}{2}mv_c^2 - mgz_0 = 0, \qquad (4.78)$$

$$\Rightarrow \quad v_c = \sqrt{2g(z_0 - r)} = \sqrt{2g}\,^6, \qquad (4.79)$$

where T_1 is the kinetic energy, U_1 the potential energy at time t_1 and E_0 the total energy at time t_0. At time t_1 the displacement of the spring-damper is zero. After the contact of the two objects they start to penetrate each other and the penalty force increases due to the displacement of the spring. At the state of maximum penetration, δ_{max} at time t_2, the position of the sphere is $z_P(t_2) = z_2 = r - \delta_{max}$ and the velocity of the sphere vanishes: $v_P(t_2) = v_2 = 0$.

In this case the spring is used to prevent further penetration of the sphere with the wall, as depicted in Fig. 4-14a). The force of the spring with stiffness c and displacement δ is given by equation (4.54). By integration of (4.54) along the path of penetration ds one obtains the potential energy of the displaced spring at maximum penetration,

$$U_s = -\int_0^{\delta_{max}} \left|\vec{F}_s(\delta)\right| ds = \frac{1}{2}c\delta_{max}^2. \qquad (4.80)$$

Evaluating the energy of the system at time t_2 leads to

$$\begin{aligned} E_2 - E_1 &= U_2 + U_s - U_1 - T_1 = 0 \\ &= mg(r - \delta_{max}) + \frac{1}{2}c\delta_{max}^2 - mgr - \frac{1}{2}mv_c^2 = 0 \end{aligned} \qquad (4.81)$$

---

[6] Note that the quantity below the square root on the right hand side of equation is in units [m·m/s²].

4 Numerical Methods

The same result can be obtained by evaluation of $E_2 - E_0$ of course. Equation (4.81) can be written as a quadratic equation for δ_{max} for a given initial impact velocity v_c, but only the positive solution is relevant,

$$\delta_{max}^2 - \frac{2mg}{c}\delta_{max} - \frac{m}{c}v_c^2 = 0 \ . \tag{4.82}$$

By inserting v_c from (4.79) into (4.82) yields

$$\delta_{max}^2 - \frac{2mg}{c}\delta_{max} - \frac{2mg}{c} = 0 \ . \tag{4.83}$$

Case B: collision of two spheres

A more complex situation is depicted in Fig. 4-14b). Assume symmetry and the sphere P collides with its mirrored counterpart P' that has identical properties but opposite velocity $v_{P'} = -v_{P'}$, and position $r' = -r$. Since gravity acts as external force for the given situation, a constant force of $2mg$ is applied to sphere P' which results in a corresponding acceleration of $g' = -g$. Furthermore, sphere P has a given velocity $v_P(t_1) = v_c$ right before contact. The systems energy for time t_1 reads

$$E_1 = U_{P,1} + T_{P,1} + U_{P',1} + T_{P',1} = 2mgr + \frac{1}{2}mv_c^2 + \frac{1}{2}m(-v_c)^2 \ . \tag{4.84}$$

For positions $z_P(t) < r$, the spheres overlap until all kinetic energy is absorbed by the spring. The potential energy of the displaced spring E_s at maximum penetration is analogous to case A (see equation (4.80)), but the spheres' potential energy at the time of maximum penetration t_2 is smaller than in the former case,

$$U_{P,2} = U_{P',2} = mg(r - \frac{1}{2}\delta_{max}) \ . \tag{4.85}$$

Evaluation of the system energy for times t_1 and t_2 leads to

$$\begin{aligned} E_2 - E_1 &= U_{P,2} + U_{P',2} + E_s - 2(U_{P,1} + T_{P,1}) \\ &= 2mg(r - \frac{1}{2}\delta_{max}) + \frac{1}{2}c\delta_{max}^2 - 2mgr - mv_c^2 = 0 \ . \end{aligned} \tag{4.86}$$

Rearranging equation (4.86) one obtains the resulting quadratic equation for δ_{max}

$$\delta_{max}^2 - \frac{2mg}{c}\delta_{max} - \frac{2m}{c}v_c^2 = 0 \tag{4.87}$$

By comparison of both cases for the same collision velocity, it is obvious that in case B the maximum penetration depth and thus also the maximum penalty force are reached faster, resulting in a shorter collision time. Hence, the interaction of a sphere with a triangle (case B) is the decisive case with regard to stability considerations as discussed later in this chapter (see section 4.3.6.2).

4.3.4 Friction

4.3.4.1 Coulomb's Law of Friction

Although friction between solid bodies is a very complicated physical phenomenon, there exists a simple law for dry friction that is an appropriate approximation for engineering applications. Based on experimental investigations, Coulomb proposed the frictional force as a function of the normal force multiplied by a friction coefficient. He distinguished two kinds of friction: static and kinetic friction.

Static Friction (also termed sticking friction)

To set a body at rest on a plane surface into motion the static friction force has to be applied,

$$\vec{F}_{Rs} = -\mu_s \left| \vec{F}_n \right| \vec{e}_t \; , \tag{4.88}$$

where μ_s is the static friction coefficient (dependent on the interacting materials but hardly on contact area or roughness) and $\vec{e}_t$ is the unit vector of the tangential component of the relative velocity perpendicular to the normal force $\vec{F}_n$.

Kinetic Friction (also termed dynamic or slipping friction)

After the force of static friction has been overcome, the resisting force of kinetic friction acts on the body. It has the same form

$$\vec{F}_{Rk} = -\mu_k \left| \vec{F}_n \right| \vec{e}_t \tag{4.89}$$

where μ_k is the kinetic friction coefficient that has similar properties as μ_s and is independent of the sliding velocity.

It may be noted, that according to Popov (2010) the static and dynamic forces are not able to be considered separately in many mechanical problems as they actually have the same origin. However, to resolve the friction process in detail and considering all interdependences with static forces at once would be very inefficient for numerical models like DEM. Hence, for such models both kinds are treated independently or in sequence depending on the application (see e.g. Brendel and Dippel (1998)).

4 Numerical Methods

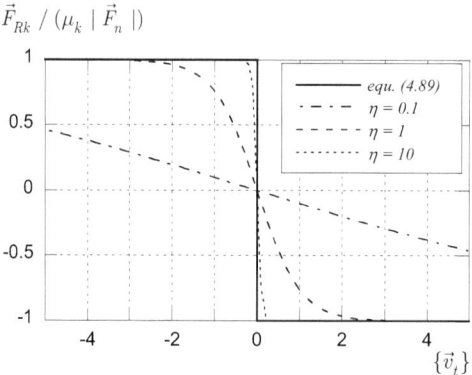

Fig. 4-15: One dimensional consideration of kinetic friction: discontinuity of the friction force at zero velocity and approximations.

One difficulty in numerical modelling of kinetic friction is the discontinuity of the friction force at zero velocity, where it changes its sign (see Fig. 4-15). Close to the discontinuity, in reality already for small tangential velocities v_t, relatively large forces occur. It may lead to numerical instabilities. To overcome this problem, the discontinuity is approximated by a continuous sigmoidal function, e.g. the hyperbolic tangent (e.g. Andersson *et al.* (2007))

$$\vec{F}_{Rk} = -\mu_k |\vec{F}_n| \tanh(\eta\{v_t\}) \vec{e}_t , \qquad (4.90)$$

where η is the slope of the function at $v_t = 0$. For different values of the friction slope η the course of the function is depicted in Fig. 4-15. If the body is a sphere, it will rather start to roll than to slide. In this case, the tangential velocity at the contact point is decisive. This velocity is called creep speed v_{cs} and is the difference between the translational velocity of the body and the circumferential velocity $\omega \delta_c$ of the sphere

$$v_{cs} = v_t - \omega \delta_c , \qquad (4.91)$$

where δ_c is the distance between the sphere centre and the contact point and $\omega = |\vec{\omega}|$ is the norm of the angular velocity of the sphere (for the case of two rotating bodies $\vec{\omega}$ is the relative angular velocity). Due to $v_{cs} < v_t$ the resulting friction force is smaller than for pure sliding, i.e. sliding without rolling. In Fig. 4-16 pure sliding and kinetic friction with rolling are compared. Both examples are for a sphere on a plane surface with acceleration due to a constant tangential force of 1 N, $|\vec{F}_n| = 1$ N and $\mu_k = 0.7$ for similar friction slopes as above. For rolling, the angular acceleration will become approximately constant after some time; this limits the creep speed to a maximum value. In the given example, the velocity of the rolling sphere at $t = 0.2$ s is approx. 1.5 m/s.

4.3 Discrete Element Method

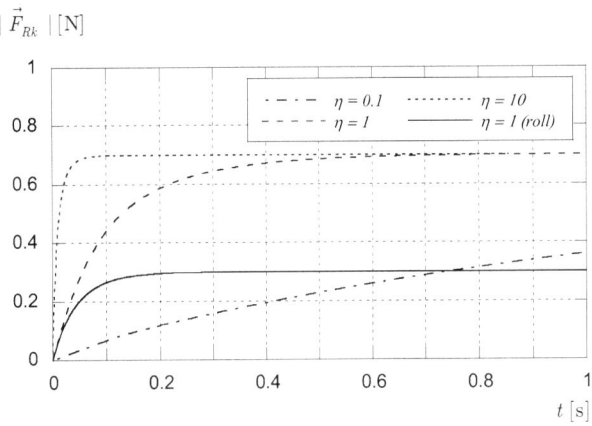

Fig. 4-16: Progression of the kinetic friction force of a body set into motion on a plane surface for different friction slopes. The special case of a sphere starting to roll is shown by the straight line.

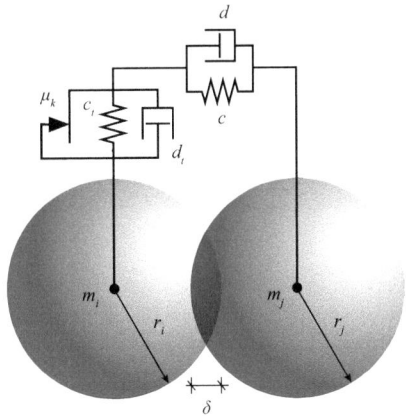

Fig. 4-17: Schematic model of interacting spheres with friction.

Static friction is more complicated to model than kinetic friction. Cundall and Strack (1979) proposed a penalty sticking friction model that inserts a tangential spring-damper system between the bodies in contact as depicted in Fig. 4-17 (note: for sphere to triangle interactions, the mirrored sphere is fixed). Thus, the bodies will actually not statically stick at the contact point but will move constrained by the spring-damper system. The elastic force of the tangential spring-damper system is

$$\vec{F}_{s,t} = -c_t \delta_t \vec{e}_t \;, \tag{4.92}$$

where c_t is the stiffness of the tangential spring, also termed tangential elasticity, and δ_t is the tangential displacement.

The maximum retaining force of the spring-damper system is given by the static friction force according to equation (4.88). Thus the behaviour of a stick friction contact can be described as

$$\frac{\left|\vec{F}_{s,t}\right|}{\left|\vec{F}_{Rs}\right|} \begin{cases} \leq 1 : \text{sticking}\,, \\ > 1 : \text{release and re-stick or slide}\,. \end{cases} \tag{4.93}$$

Damping of the system is introduced according to equation (4.59) and controlled by the attenuation factor γ_d via

$$d_t = 2\gamma_d \sqrt{mc_t} \;, \tag{4.94}$$

where m is the mass of the heavier of the two interacting bodies. The resulting dissipative force of the stick friction contact can be written as

$$\vec{F}_{D,t} = -\left(c_t \delta_t + d_t v_t\right)\vec{e}_t \;. \tag{4.95}$$

The main purpose of including damping in a static friction contact is to prevent oscillations of the interacting bodies if the acting force is smaller than the retaining force. Furthermore, damping leads to a slower displacement of the spring, i.e. it takes longer to reach the maximum retaining force and thus to a delayed release of the contact.

An example, rather illustrative than realistic, of a pure static friction contact is shown in Fig. 4-18 with constant tangential force of 1 N, $\left|\vec{F}_n\right|$ = 1 N, μ_s = 0.7 and for different values of d_t as well as c_t.

As mentioned above, the case of pure stick friction for a body with constant acceleration as depicted in Fig. 4-18 is not very realistic. On the one hand this is so because the body does not return to a state of no motion and the static friction is permanently acting. On the other hand this is so because according to Coulomb's model stick friction is an instantaneous force, i.e. the stiffness of the tangential spring should be large, e.g. $c_t = 10^6$ N/m.

The combination of static and kinetic friction is called stick-slip friction, i.e. sticking and slipping occur in sequence.

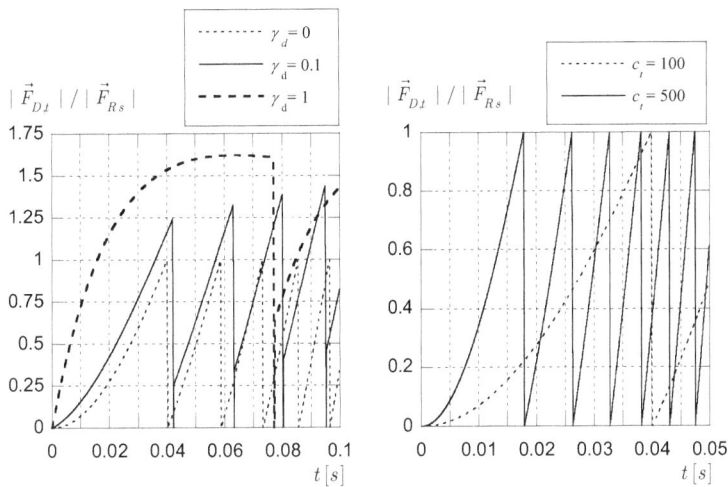

Fig. 4-18: Pure static friction for different attenuation factors and tangential elasticities. Left: $c_t = 100$, right: $\gamma_d = 0$

4.3.4.2 Rolling Resistance

According to Popov (2010), stick-slip friction is the basic process at the contact zone when rolling occurs. However, pure slip friction is already a simple approach for rolling resistance, since a sphere will not roll without an eccentrically acting force causing a torque. To account for the rather complex process of rolling resistance, a friction torque may in addition be considered according to Zhou et al. (1999). By application of the same considerations as for slip friction the friction torque reads

$$\vec{M}_f = -\mu_r \left|\vec{F}_n\right| \tanh\left(\eta\{|\vec{\omega}|\}\right) \frac{\vec{\omega}}{|\vec{\omega}|} \,, \qquad (4.96)$$

where μ_r is the rolling resistance coefficient with units [m] and $\vec{\omega}$ is the angular velocity (for two rotating bodies $\vec{\omega}$ is the relative angular velocity). Quoted numerical values for μ_r are in the range of 0.0002 to 0.015 m. In this work, the spheres are considered to be rigid, hence the choice of $\mu_r = 0.001$ m seems to be reasonable. The impact of rolling friction is depicted in Fig. 4-19 for a constant tangential force of 1 N, $\vec{F}_n = 1$ N, $\mu_k = 0.7$, $\eta = 10$ s/m and $\mu_r = 0.001$ m.

The friction models discussed here are considered to be suitable for modelling the basic stick, slip and rolling friction interactions between spherical particles and are numerically stable if appropriate parameters are chosen. Some advanced friction models are presented by Andersson et al. (2007).

4 Numerical Methods

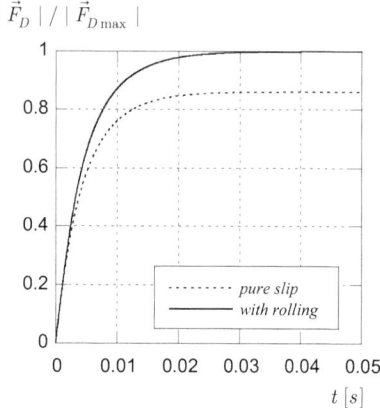

Fig. 4-19: Comparison of pure slip friction and friction with additional rolling resistance, where $|\vec{F}_D|$ and $|\vec{F}_{D\max}|$ are the total dissipative force and its maximum, respectively ($\mu_k = 0.7$, $\mu_r = 0.001\,m$, $\eta = 10$).

4.3.5 Time Integration

4.3.5.1 Integration Scheme

A system of rigid bodies that is simulated by the introduced models can be advanced in time by solution of the equations of motion, (3.28) and (3.31), with an appropriate scheme for time integration. The main goal of the time integration is to resolve accurately every collision occurring in an interval (i.e. during a discrete time step Δt) and to preserve stability. The first is fulfilled, if the size of the time step will not be larger than the smallest diameter of a particle divided by the maximum occurring velocity; this is comparable to the CFL condition in computational fluid dynamics (see section 4.2.7.2). The latter is quite challenging since the contact forces are generally non-smooth and large discontinuities may arise at the right hand sides of the equations of motion.

One of the most common time integration schemes in molecular dynamics is the integrator presented by Verlet (1967). The scheme is reversible and quite simple,

$$\vec{r}_i(t+\Delta t) = 2\vec{r}_i(t) - \vec{r}_i(t-\Delta t) + \Delta t^2 \vec{a}_i \,, \tag{4.97}$$

where the two first terms on the right-hand side describe the translation of the particle by constant velocity and the last term its correction due to applied forces. The velocity can be evaluated by the finite difference term

$$\vec{v}_i(t) = \frac{\vec{r}_i(t+\Delta t) - \vec{r}_i(t-\Delta t)}{2\Delta t} \,. \tag{4.98}$$

Note that the velocity is one step behind. To overcome this drawback, derivatives of the Verlet integrator are available, such as the so-called Velocity-Verlet scheme where positions, velocities and accelerations at time $t+\Delta t$ are estimated based on their values at time t. Another similar

4.3 Discrete Element Method

approach is the leap-frog algorithm, where the position depends on intermediate values of the velocity (see e.g. Fincham (1992)), that can also be used in combination with a predictor-corrector scheme (for a PC-leapfrog scheme see section 4.2.7.1). These schemes can be modified to be used for the time integration of rotational dynamics with quaternions, as proposed e.g. by Omelyan (1998). However, all Verlet-derivates are only conditionally stable.

According to Fleissner (2010), another group of time integration routines was found to be more suitable for applications of a dissipative DEM as used in this work. He proposes the use of the Newmark-β methods which range from fully explicit to fully implicit depending on parameters β and γ, i.e. for the former $\beta = \gamma = 0$ and for the latter $\beta = 1/2, \gamma = 1$. For intermediate values $\beta = 1/4$ and $\gamma = 1/2$, an unconditionally stable implicit scheme results that is of second order accuracy and has negligible numerical damping. For the use with particle systems, the implicit schemes can be solved by a predictor-corrector method. Furthermore, Fleissner and co-workers extended the scheme by a modification of Omelyan's approach to be used for the integration of rotational dynamics.

4.3.5.2 Time-step size

By application of implicit, unconditionally stable schemes for time integration such as the Newmark-β methods, there is per se no correlation between time-step size and stability. The size of the overall time step actually depends on the desired accuracy. This may not go hand in hand with little computational efforts and a fast simulation progress. Therefore, use of the largest possible time step that meets the accuracy requirements is desired. This can be achieved by time step control algorithms that dynamically adjust the time step for a corresponding state of the system. Fleissner (2010) successfully adopted the approach by Zohdi (2005) for Newmark-β methods and the Generalized-α methods (not discussed herein).

For the explicit PC-leapfrog scheme, the size of a time step can be obtained by similar conditions as for the SPH method. The relevant length scale is the radius of the smallest sphere $r_{\min}$ and the maximum occurring velocity is taken as the reference velocity, i.e. $u_{ref} = |\vec{u}|_{\max}$. Furthermore, viscous forces are not considered. Including a safety factor α_s, this leads to the following conditions for the time-step size

$$\Delta t = \alpha_s \min\left(\frac{r_{\min}}{u_{ref}}, \sqrt{\frac{r_{\min}}{a_{\max}}}\right). \tag{4.99}$$

4.3.5.3 Solution Algorithm

The solution algorithm for the DEM can be outlined as follows:

I. Set initial conditions
II. While time $t \leq t_{end}$
  1. set boundary conditions,
  2. determine time-step size Δt according to section 4.3.5.2,
  3. for each particle i:
    3.1. determine interacting particles j and compute applied forces $\vec{F}_{a,i}$ and torques $\vec{M}_{a,i}$ according to equations (3.32) and (3.33).
    3.2. solve equations (3.28), (3.31), (3.36) and (3.37) with an appropriate time integration scheme (see 4.3.5.1), resulting in temporally advanced translational and rotational state of particle i.
  4. advance particles to their new positions and increment time by Δt.

For a detailed description of the simulation algorithm including the handling of discrete events please refer to Fleissner (2010).

4.3.6 Considerations of Accuracy

4.3.6.1 Contact Time

From a physical point of view, the error of the method depends on how precisely the maximum penetration can be reached, i.e. how well the impact and exerted penalty forces are balanced. This implies a theoretical maximum size of the time step that is half the contact time or a fraction of it for real model conditions. An estimate for the impact time may be derived from an equivalent model of a harmonic oscillator, i.e. the path of impact can be regarded as a sectional motion of a corresponding harmonic oscillator (see Fig. 4-20). Therefore, equation (4.53) for the collision of two identical spheres as depicted in Fig. 4-8 with the linear force law, without damping and without constraint according to equation (4.47) is considered

$$\ddot{\zeta} + \omega_0^2 \zeta = 0 \ . \tag{4.100}$$

This equation is similar to (4.60) with $\gamma = 0$. For a harmonic oscillator with only one mass as in Fig. 4-14a) the mass is $\mu_m = m$ (sphere-sphere interaction) and for a harmonic oscillator with two masses as in Fig. 4-14b) its mass has to be reduced according to (4.51) (sphere-triangle interaction). The solution to (4.100) is

$$\zeta = A_o \cos(\omega_0 t + \theta) \ . \tag{4.101}$$

The maximum penetration $\delta_{\max}$ of two identical spheres with impact velocity v_c is given by (4.87). For the derivation of the impact time, only the penetration path of one sphere is of interest. Thus, the maximum displacement of one sphere is $\hat{\delta}_{\max} = \delta_{\max}/2$. Furthermore, the equilibrium position

4.3 Discrete Element Method

δ_0 can be obtained from equation (4.56). As introduced in section 4.3.2.1, a harmonic oscillator will move around the equilibrium position, leading to an amplitude $|A_o| = \hat{\delta}_{max} - \delta_0$.

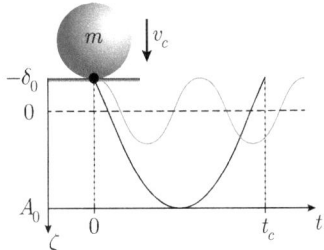

Fig. 4-20: Definition sketch for the equivalent model of a harmonic oscillator. The paths of the oscillator with zero impact velocity (grey line) and the equivalent model with impact velocity v_c are shown.

The other parameters of equation (4.101) can be found with appropriate initial conditions for the problem of interest. At the time of impact $t = 0$ s, the impact velocity is $v_c = |\vec{v}_c|$ and the offset from the equilibrium position is $-\delta_0$ and the corresponding amplitude is $A_o = -(\hat{\delta}_{max} - \delta_0)$; thus the initial conditions are

$$\zeta(0) = -\delta_0 , \qquad (4.102)$$

$$v(0) = v_c . \qquad (4.103)$$

The corresponding phase angle (in radians) can be obtained by evaluation of equation (4.101) with initial condition (4.102); the resulting phase angle is

$$\theta = \arccos\left(\frac{\delta_0}{\hat{\delta}_{max} - \delta_0}\right) . \qquad (4.104)$$

Evaluating the time derivative of ζ given in equation (4.101) for the second initial condition (4.103) leads to

$$v(0) = \left(\frac{d\zeta}{dt}\right)\bigg|_{t=0} = A_o \omega_0 \sin(\theta) = v_c , \qquad (4.105)$$

where $\omega_0 = 2\pi/T_o$ and T_o [s] is the period of the harmonic oscillator. Inserting the phase angle θ, angular frequency ω_0 and amplitude A_o into equation (4.105), one obtains by rearranging

$$T_o = \frac{(\hat{\delta}_{max} - \delta_0) 2\pi \sin(\theta)}{v_c} , \qquad (4.106)$$

which is the period of the equivalent model. This result is similar to the simpler form of equation (4.61), since ω_0 is constant for given mass and stiffness and hence T_o is constant as well. Nevertheless, equation (4.106) can be used to obtain an expression for the phase angle independent of δ_0 and $\hat{\delta}_{max}$. By rearranging equations (4.106) and substituting (4.104) into expressions for

$\sin(\theta)$ and $\cos(\theta)$, as well as inserting equation (4.56) for δ_0, the tangent of the phase angle can be written as

$$\tan(\theta) = \frac{v_c}{g}\sqrt{\frac{c}{\mu_m}} = \frac{v_c}{g}\omega_0. \qquad (4.107)$$

According to the initial condition (4.102), the impact starts at offset δ_0, corresponding to a phase angle θ, and it will end at $2\pi - \theta$. Thus the collision time t_c is $2(\pi - \theta)/\omega_0$ or by inserting $\omega_0 = 2\pi/T_o$

$$t_c = T_o\left(1 - \frac{\theta}{\pi}\right) \qquad (4.108)$$

with period $T_o = 2\pi\sqrt{\mu_m/c}$.

Equation (4.108) allows for a reasonable estimation of the contact time of a collision modelled by the linear force law. With regard to the similarity considerations made in section 4.3.2.2, i.e. by choice of the stiffness according to equation (4.70), this approach may be used for the Hertz force model as well.

4.3.6.2 Error Depending on Size of Time Step

As introduced in the previous section, the choice of an appropriate size of the simulation time step is important for a correct capturing of the impact process. If the time step is too large, the maximum penetration that corresponds to equilibrium conditions may be exceeded, and linear momentum will not be conserved. To prevent such situations, the size of the time step Δt is usually chosen to be much smaller than the contact time t_c, i.e. $\Delta t \ll t_c/2$. Note that the effective maximum simulation time step is a property of the time integration scheme and is based on numerical considerations as introduced in section 4.3.5.2. Thus the resulting maximum size of the time step does not necessarily avoid situations as mentioned above, and the accuracy of the simulation depends still strongly on the kind of scheme as well as on the choice of appropriate parameters. This can be illustrated by the following example where the PC-leapfrog integration has been used. Consider a similar setup as depicted in Fig. 4-14, with a sphere of diameter $d_s = 0.03$ m, density $\rho_s = 2800$ kg/m^3 and impact velocity $v_c = 0.16$ m/s. The material of the sphere is granite ($E_k = 60 \times 10^9$ N/m^2, $\nu_k = 0.25$) and the stiffness of the linear law is obtained from equation (4.68), i.e. an equal maximum penetration depth is assumed for comparison of the penalty force laws.

4.3 Discrete Element Method

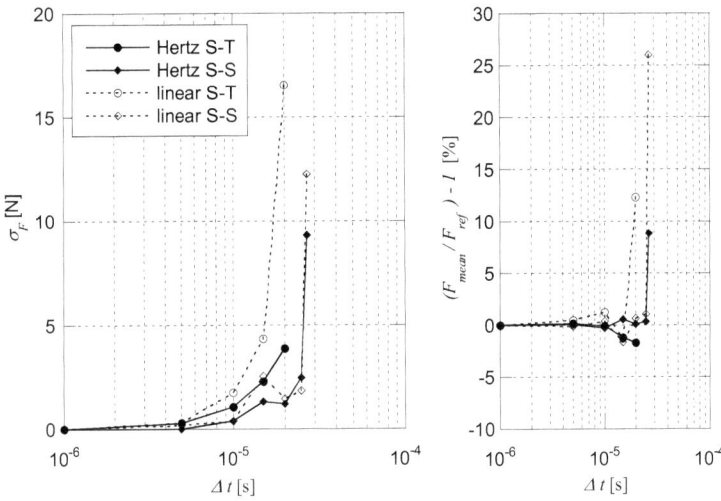

Fig. 4-21: Comparison of linear and Hertz law for different time-step sizes as well as for sphere to sphere (S-S) and sphere to triangle (S-T) collisions.

The error in terms of the standard deviation σ_F of the penalty force depending on the simulation time step Δt are depicted in Fig. 4-21 for sphere-sphere (S-S) and sphere-triangle (S-T) interactions with the penalty force according to the linear and the Hertz laws. Moreover, the ratio of the mean penalty force F_{mean} and the reference penalty force F_{ref} that corresponds to equilibrium conditions is shown. Furthermore, the ratio $t_c/2\Delta t$, i.e. the number of time steps used to achieve maximum penetration, is diagrammed in Fig. 4-22 for the given example. As can be seen, for the chosen realistic model parameters the error as well as the deviation of the mean penalty force rapidly increases for a time-step size larger than 10^{-5} s. This is not particularly remarkable since for the given case a time step $\Delta t \geq 10^{-5}$ s is quite close to half the interaction time, i.e. $t_c/2\Delta t \leq 10$.

The conclusion made in the last paragraph of section 4.3.3, "that the sphere-triangle interaction is the relevant case with regard to stability considerations" is confirmed by the above presented results. For the S-T interaction, the error increases faster and the maximum time-step size is smaller than for sphere-sphere interaction. This is due to its shorter collision time as shown in Fig. 4-22. Furthermore, the difference between the two force laws may be pointed out. With regard to the standard deviation of the penalty force, the Hertz force law seems to be more docile than the linear law for the same configuration, as can be seen from Fig. 4-21. The reason for this is the larger penalty force of the nonlinear Hertz law for penetration ratios δ/δ_{max} greater than approximately 0.65 (see Fig. 4-10).

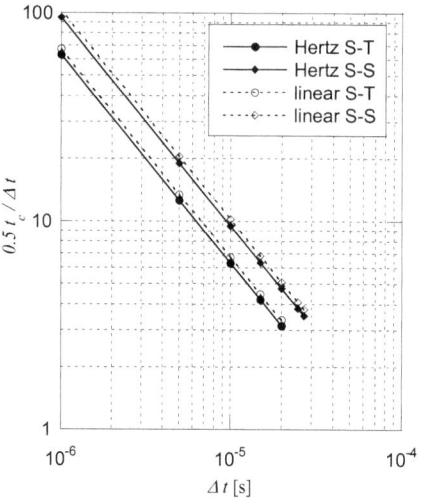

Fig. 4-22: Number of time steps used to achieve maximum penetration for the given example. The lower ends of the lines indicate that the maximum time-step size has been reached.

4.3.7 Choice of Appropriate Simulation Models and Parameters

4.3.7.1 Force Model

From the physical point of view, linear or Hertz laws are most suitable for modelling collisions of rigid spheres. Based on the insights gained in the previous section, use of the Hertz law is preferred. Moreover, the parameters of the Hertz law, i.e. Poisson's ratio ν_k and Young's modulus E_k, are material properties commonly used in engineering practice. However, this connection to real materials has to be used with care, since the emphasis in this work is on the interaction forces between rigid spheres and the corresponding conservation of momentum and not on the contact of real material in detail. Although, the Hertz law is a reasonable model for the latter, the size of the time step would be very small already for moderate accuracy when it comes to very stiff materials such as granite (see previous section). Thus, use of modified material properties which allow for larger time steps while still maintaining the accuracy requirements seems to be a useful approach. However, in such a case a larger penetration, i.e. a larger displacement of the sphere, has to be accepted.

This approach is investigated for the preferred Hertz law and the relevant sphere-triangle interaction, with the same parameters as used for the example in the previous section except that Young's modulus is varied. Consider a maximum penetration $\delta_{\max}$, or the ratio of $\delta_{\max}$ and the sphere radius

$$a = \delta_{\max}/r_s \qquad (4.109)$$

4.3 Discrete Element Method

as a measure of displacement. For the given model parameters, the sphere mass m_s and estimated impact velocity v_c, a corresponding stiffness can be obtained by rearranging equation (4.87):

$$c = \frac{2m_s}{\delta_{\max}}\left(g\delta_{\max} + v_c^2\right). \qquad (4.110)$$

The material parameter K for the Hertz law can be determined by rearranging equation (4.68):

$$K = \frac{5}{4}\frac{c}{\sqrt{\delta_{\max}}}. \qquad (4.111)$$

For a given Poisson ratio ν_k, the corresponding Young's moduli E_k can be obtained by combination of equations (4.64) and (4.65) for two identical spheres. This yields

$$E_k = \frac{3K\left(1 - \nu_k^2\right)}{\sqrt{2r_s}}. \qquad (4.112)$$

Finally, the contact time can be estimated by evaluation of equations (4.70), (4.107) and (4.108). The numerical results for different Young's moduli are summarized in Tab. 4-1.

Tab. 4-1: *Numerical results of S-T interaction with Hertz' law for different Young's moduli E_k, where $\delta_{\max}$ is the maximum penetration, t_c is the contact time and r_s the sphere radius.*

$\delta_{\max}$ [m]	$a = \delta_{\max}/r_s$ [%]	E_k [N/m²]	t_c [s]
1.37E-05	0.091	6.00E+10	1.26E-04
1.50E-05	0.1	4.75E+10	1.38E-04
7.50E-05	0.5	8.69E+08	6.86E-04
1.50E-04	1	1.58E+08	1.36E-03
3.00E-04	2	2.94E+07	2.70E-03
7.50E-04	5	3.44E+06	6.52E-03

The error in terms of the standard deviation of the penalty force which depends on the time-step size for the cases listed in Tab. 4-1 is shown in Fig. 4-23. For the given example, a ratio of $a = 1\ \%$ and a maximum standard deviation of $\sigma_F = 1$ N results for a time-step size of approximately $\Delta t \approx 3 \cdot 10^{-4}$ s; this is one order of magnitude larger than for real material conditions ($a = 0.09\ \%$).

The procedure outlined above, i.e. equations (4.109) to (4.112) in combination with Fig. 4-23, can be used to estimate the time-step size for a desired accuracy in terms of the standard deviation of the penalty force and given model parameters. If only sphere-sphere interactions are considered, the standard deviations will be approximately half of those for sphere-triangle interactions (compare Fig. 4-21).

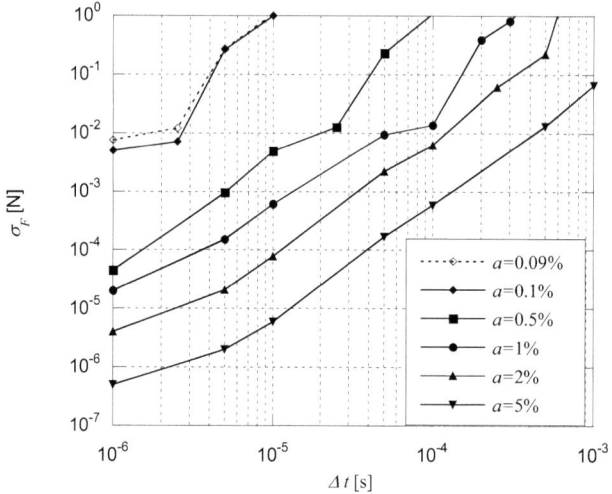

Fig. 4-23: Standard deviation of the penalty force depending on the time-step size for the cases listed in Tab. 4-1.

4.3.7.2 Stiffness

If both spheres have identical positions, the displacement of the spring is maximum, i.e. $\delta = (r_i + r_j)$. If interpenetration is not intended, the stiffness has to be chosen larger than the minimum stiffness $c_{\min}$. Based on equation (4.87) the minimum stiffness for which the two spheres do not interpenetrate can be determined by the formula

$$c_{\min} = \frac{m_{\max}(v_{\max}^2 + 2gr_{\max})}{2r_{\max}^2}, \qquad (4.113)$$

or when expressed in terms of the density ρ

$$c_{\min} = \frac{2}{3}\pi \rho r_{\max}(v_{\max}^2 + 2gr_{\max}) . \qquad (4.114)$$

4.3.7.3 Damping

The introduction of damping is reasonable to avoid high frequency oscillations that may enforce small time steps (see Gaugele *et al.* (2008)). Furthermore, it is a simple approach to account for the discrepancy of the collision force for non-uniformly shaped particles, such as real gravel and spherical particles as considered in this work.

4.4 Fluid-Structure Interaction

The interaction between fluid and structures, such as spheres or bodies with a triangulated surface, can be modelled in a similar way as the interaction of two spheres or a sphere with a triangle, respectively (see chapter 4.3.3). Therefore, in a collision of a fluid with a triangulated surface, e.g. a wall, the mirrored boundary particle is a fluid particle. Different laws for the penalty as well as the friction force at the interaction boundary are presented below. Some alternative approaches found in the literature are briefly discussed in chapter 4.2.6.3.

To determine the state of the interaction of a fluid particle with a rigid boundary, i.e. a triangle or a sphere, a condition similar to (4.47) is introduced:

$$\delta = \left(h + r_i\right) - r_{ij} \quad \begin{cases} \leq 0 : inactive, \\ > 0 : active. \end{cases} \quad (4.115)$$

Here h is the smoothing length of the SPH fluid particle and r_{ij} is the distance between the two particles. If the interacting partner is a sphere, r_i is equal to the radius of the sphere. For interaction with a triangle, $r_i = 0$ and r_{ij} is the distance of the fluid particle normal to the triangle surface.

4.4.1 Normal Forces

4.4.1.1 Modified Lennard-Jones Potential

To model the interaction of fluid particles with a rigid body in the same manner as molecular interaction seems to be a reasonable approach. In chapter 4.3.2.4, the force law for interacting molecules, the Lennard-Jones (LJ) potential, has been introduced. Monaghan (1994) used a LJ potential as wall boundary condition for simulating free surface flows (for details see appendix A.4.2). Unlike the original form, Monaghan defined the potential to be positive for repulsive forces and negative for attraction, i.e. for the exponents $m > n$. He noticed, that the approach by Peskin (1977) using a delta function as force law is another way to model rigid boundaries, but in his work the use of forces based on known molecular forces produced better results.

A further modification of the use of the LJ potential with SPH is suggested by Muller *et al.* (2004) and may be called Modified Lennart-Jones (MLJ) potential. Other than the original LJ potential that leads to an infinitely large force for a particle distance towards zero ($\left|\vec{F}_n(r_{ij} \to 0)\right| \to \infty$), they propose a force law with a finite value of the force at the boundary ($\left|\vec{F}_n(r_{ij} = 0)\right| = k$)

$$\vec{F}_n(r_{ij}) = \begin{cases} \dfrac{k}{R^2(2R - r_0)r_0}\left((R - r_{ij})^4 - (R - r_0)^2(R - r_{ij})^2\right)\vec{e}_{ij}, & if\ r_{ij} < R, \\ 0 & otherwise. \end{cases} \quad (4.116)$$

The maximum force value k at the boundary may be also denoted as stiffness of the boundary. Furthermore, the influence of the potential is limited to a given distance R, i.e. $\left|\vec{F}_n(r_{ij})\right| = 0$ for $r_{ij} \geq R$. The point where the force changes from repulsion to attraction, i.e. $\left|\vec{F}_n(r_{ij})\right| = 0$, can be

4 Numerical Methods

set equal to r_0. The authors argue that this approach is important for numerical robustness, especially for real time simulations. See appendix A.4.3 for further details.

For the investigation of wall bounded flows, an approach depending on the particle distance to the boundary δ_w is preferable. For the interaction of a fluid particle with a triangle the following transformation is defined:

$$fluid - triangle: \quad \delta_w := r_{ij}/2, \delta_{w0} := r_0/2, D_w := R/2 \ . \tag{4.117}$$

In what follows, for the sake of clarity, only relations for the fluid particle to triangle interaction are discussed. In appendix A.3 the transformation and derived equations for the interaction of a fluid particle with a sphere are summarised for later reference. By insertion of (4.117) into (4.116) one obtains

$$\vec{F}_n(\delta_w) = \begin{cases} \dfrac{k}{D_w^2(2D_w - \delta_{w0})\delta_{w0}} \left((D_w - \delta_w)^4 - (D_w - \delta_{w0})^2(D_w - \delta_w)^2\right)\vec{e}_{ij}, & \text{if } \delta_w < D_w, \\ 0 & \text{otherwise} \end{cases} \tag{4.118}$$

and the corresponding potential reads

$$U(\delta_w) = \frac{k}{D_w^2(2D_w - \delta_{w0})\delta_{w0}} \left[\frac{\delta_w^5}{5} - D_w \delta_w^4 + D_w^2 \delta_w (2D_w - \delta_{w0})\delta_{w0} \right. \\
\left. - D_w \delta_w^2 (D_w^2 + 2D_w \delta_{w0} - \delta_{w0}^2) \right. \\
\left. + \frac{\delta_w^3}{3}(5D_w^2 + 2D_w \delta_{w0} - \delta_{w0}^2) \right], \tag{4.119}$$

$$\text{if } \delta_w < D_w, \text{ otherwise } U(\delta_w) = 0 \ .$$

An example of the progression of the MLJ potential and force for $\delta_{w0}/D_w = 0.2$ and $k = 100$ N is depicted in Fig. 4-24.

When considering only repulsive forces, i.e. $D_w = \delta_{w0}$, equation (4.118) becomes

$$\vec{F}_n(\delta_w) = \frac{k}{\delta_{w0}^4}(\delta_{w0} - \delta_w)^4 \vec{e}_{ij}, \quad \delta_w \leq \delta_{w0} \ . \tag{4.120}$$

The corresponding potential reads

$$U(\delta_w) = \frac{k}{\delta_{w0}^4} \left(\frac{\delta_w^5}{5} - \delta_w^4 \delta_{w0} + 2\delta_w^3 \delta_{w0}^2 - 2\delta_w^2 \delta_{w0}^3 + \delta_w \delta_{w0}^4 \right), \quad \delta_w \leq \delta_{w0} \ . \tag{4.121}$$

4.4 Fluid-Structure Interaction

The maximum of the exclusively repulsive potential is at distance δ_{w0} from the wall and has the value

$$U(\delta_{w0}) = \frac{1}{5}k\delta_{w0} \, . \tag{4.122}$$

The progression of a repulsive MLJ potential and force is depicted in Fig. 4-25.

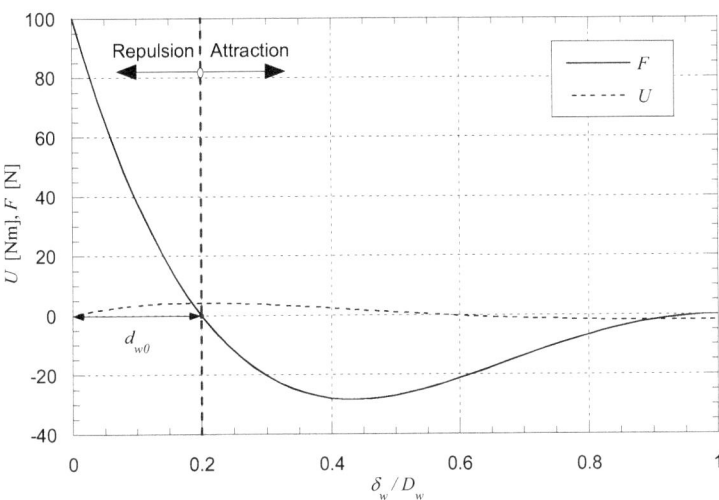

Fig. 4-24: Behaviour of the modified Lennard-Jones potential for parameters $\delta_{w0}/D_w = 0.2$ and $k = 100$ N and $F = \left|\vec{F}_n(d_w)\right|$.

To balance a given external static force of amount F, e.g. the weight mg of a fluid body with mass m, the equilibrium distance to the wall can be determined by

$$\delta_{weq} = \delta_{w0}\left(1 - \left(\frac{F}{k}\right)^{1/4}\right), \tag{4.123}$$

and the corresponding potential is

$$U(\delta_{weq}) = \frac{\delta_{w0}}{5}\left(k - F\left(\frac{F}{k}\right)^{1/4}\right). \tag{4.124}$$

If a certain equilibrium distance to the wall is preferred, the appropriate stiffness could be obtained by rearranging (4.123),

$$k = F\left(\frac{\delta_{w0}}{\delta_{w0} - \delta_{weq}}\right)^4. \tag{4.125}$$

The situation of a SPH particle close to a wall at an equilibrium position is depicted in Fig. 4-26.

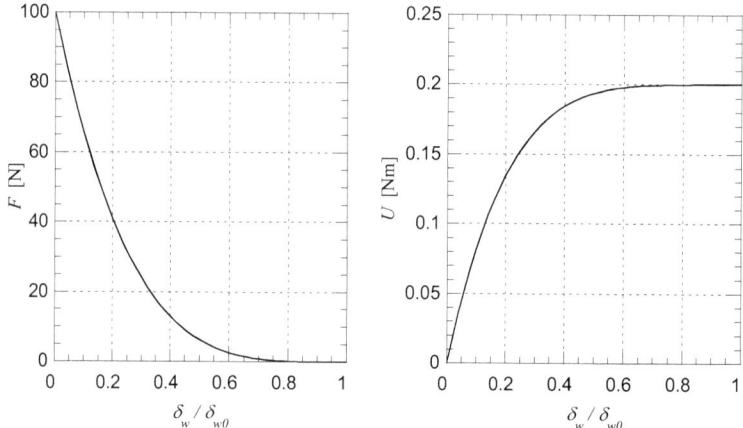

Fig. 4-25: Behaviour close to the wall of an exclusively repulsive modified Lennard-Jones potential with $k = 100\ N$ and $F = \left|\vec{F}_n(d_w)\right|$.

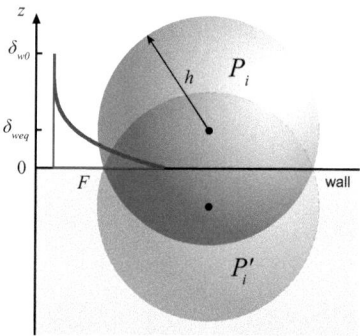

Fig. 4-26: SPH particle P_i with smoothing length h close to a wall at equilibrium position δ_{weq}. The counter particle P_i' is used to obtain appropriate boundary conditions.

The minimum stiffness k_{min} of the boundary for non-static conditions depends on the dynamic load that can be expressed by its total energy at the point δ_{w0} where the interaction starts,

$$E_{tot}(\delta_{w0}) = mg\delta_{w0} + \frac{1}{2}mu_w^2 , \qquad (4.126)$$

where m is the mass of an approaching body and u_w its velocity perpendicular to the boundary. This energy has to be balanced with the MLJ potential if no penetration is desired. Combining (4.126) and (4.122) yields

$$k_{min} = 5m\left(g + \frac{u_w^2}{2\delta_{w0}}\right) . \qquad (4.127)$$

For a potential with $\delta_{weq} = \delta_{w0}/2$, equations (4.125) and (4.124) reduce to

$$k = 16F; \quad F = \frac{k}{16} , \qquad (4.128)$$

$$U(\delta_{weq}) = \frac{31}{160}k\delta_{w0} = \frac{31}{10}F\delta_{w0} . \qquad (4.129)$$

4.4.1.2 Force Based on Kernel Gradient

In the previous approach the boundary collision is modelled as an interaction of a particle with its mirrored counterpart and penetration was prevented by a force based on a MLJ potential. Another approach for modelling the interaction of a fluid particle with a rigid body was introduced by Monaghan et al. (2003). They propose to use a fixed boundary particle with given mass and density plus a kernel function. Thus, the repulsive force may act as that between a pair of fluid particles. Furthermore, they also consider particles tangential to the boundary for the calculation of the boundary force.

The original approach has been adapted by Lehnart (2008) in the context of her work for normal interaction with the boundary particle only, i.e. without taking tangential boundary particles into account. This approach is restricted to the interaction of fluid particles with triangles. The corresponding boundary force reads

$$\vec{F}_n(\delta_w) = \left(\frac{\alpha_g}{1+\alpha_g}\right)m_w\Gamma_g(\delta_w)\vec{e}_{ij} , \qquad (4.130)$$

where $\alpha_g = \rho_f/\rho_w$ is the ratio of the fluid density ρ_f and the density of the boundary particle ρ_w and m_w is its mass. The function $\Gamma_g(\delta_w)$ has the form of the gradient of a cubic spline kernel and is defined in terms of $q_g = \delta_w/h$ as

4 Numerical Methods

$$\Gamma_g(\delta_w) = \beta_k \begin{cases} \frac{2}{3} & \text{if } 0 < q_g \leq \frac{2}{3}, \\ 2q_g - \frac{3}{2}q_g^2 & \text{if } \frac{2}{3} < q_g \leq 1, \\ \frac{1}{2}(2 - q_g)^2 & \text{if } 1 < q_g < 2, \\ 0 & \text{otherwise}, \end{cases} \qquad (4.131)$$

where $\beta_k = 0.02 c_s^2 / \delta_w$, and c_s is the sound velocity, see Monaghan et al. (2003).

To balance a given external static force of amount F at distance δ_{weq} from the boundary, the density of the boundary particle can be determined by

$$\rho_w(\delta_{weq}) = -\frac{F\rho_f}{F - m_w \Gamma_g(\delta_{weq})}. \qquad (4.132)$$

For hydrostatic pressure conditions, the choice of $\rho_f = 1005$ kg/m³ is recommended, and the mass m_w of the boundary particle can be approximated by the average mass of a fluid particle.

4.4.2 Friction

4.4.2.1 Tangential Force

The friction between the fluid and the surface of a sphere or triangle can be modelled in a similar way as the friction between solid bodies as described in chapter 4.3.4. However, the exerted tangential force is actually a viscous shear force plus effects due to the character of the surface. Thus, the friction coefficient depends rather on the viscosity of the fluid and the surface roughness than on dry material-to-material properties.

With respect to the above introduced approaches for the boundary normal force $\vec{F}_n(\delta_w)$ the viscous friction force can be written as

$$\vec{F}_{Rv}(\delta_w) = -\mu_v \left| \vec{F}_n(\delta_w) \right| \tanh\left(\eta\{v_t\}\right) \vec{e}_t, \qquad (4.133)$$

where μ_v [-] is the coefficient of viscous friction. The hyperbolic tangent of $\eta\{v_t\}$, where η is the friction slope and $\{v_t\}$ is the value of relative tangential velocity, has the purpose to smooth the discontinuous behaviour analogue to the approach for kinetic friction (compare equation (4.90)).

4.4.2.2 Terminal Velocity

Since the hyperbolic tangent has a maximum value of 1 that is approximately reached for arguments equal to or larger than approx. 4, the maximum of the viscous friction force is given by

4.4 Fluid-Structure Interaction

$$F_{Rw}(\delta_w)_{\max} = \mu_v \left| \vec{F}_n(\delta_w) \right| . \tag{4.134}$$

As for kinetic friction, viscous friction only occurs if a tangential force is applied that acts in opposite direction of the friction force, i.e. $\{v_t\} \neq 0$. Furthermore, the exerted friction force cannot be larger than the applied force. Consider an applied tangential force of amount F_t acting on a fluid particle for which the ratio between F_t and the maximum viscous friction force is defined as

$$\psi_R = \frac{F_t}{F_{Rw}(\delta_w)_{\max}} . \tag{4.135}$$

Thus, for $\psi_R > 1$ a fluid particle of mass m will be continuously accelerated by the amount $\left(F_t - F_{Rw}(\delta_w)_{\max}\right)/m$, and for $\psi_R \leq 1$ the friction force will be limited to

$$F_t = F_{Rw}(\delta_w)_{\max} \tanh\left(\eta\{v_t\}\right) . \tag{4.136}$$

Rearranging equation (4.136) and introducing ψ_R leads to the terminal velocity of the fluid particle for $\psi_R \leq 1$ in the form

$$\{v_{end}\} = \frac{1}{\eta} \operatorname{atanh}(\psi_R), \quad \psi_R \leq 1 . \tag{4.137}$$

Furthermore, equation (4.136) can be rearranged to obtain a viscous friction coefficient that corresponds to a desired tangential velocity v_{end} and a given friction slope η,

$$\mu_v(\eta, v_{end}) = \frac{F_t}{\left|\vec{F}_n(\delta_w)\right| \tanh\left(\eta\{v_{end}\}\right)}, \quad \psi_R \leq 1 . \tag{4.138}$$

Rearranging equation (4.138) with $\psi_R = 1$ leads to the minimum friction coefficient that is necessary to obtain a desired terminal velocity for given forces $\vec{F}_t$ and $\vec{F}_n$

$$\mu_{v,\min} = \frac{\left|\vec{F}_t\right|}{\left|\vec{F}_n\right|} . \tag{4.139}$$

For example, consider a tangential force of amount $F_t = 5$ N, a normal force of amount $\left|\vec{F}_n(\delta_w)\right| = 50$ N and a constant tangential acceleration of $a_t = 1$ m/s^2. The friction slope is then given as $\eta = 0.1/\{v_{end}\}$, i.e. $\eta\{v_{end}\} = 0.1$, yielding a viscous friction coefficient $\mu_v \approx 1.0033$. Various combinations of η and v_t are depicted in Fig. 4-27.

Note that the acceleration of a particle under the influence of friction is controlled by the friction slope and is generally limited by the applied force as indicated by the straight line "no friction" in Fig. 4-27. However, bear in mind that the terminal velocity may be reached fastest with small values of η because of little resistance due to friction at the beginning. By contrast, large values of

4 Numerical Methods

η may lead to a rapid increase of friction for small velocities already. Large values of the friction slope η, e.g. $\eta = 4/\{v_t\}$ corresponding to $\tanh(4) \approx 1$, are not preferred since there will be only a small temporal lag in the build-up of the friction force, possibly leading to undesired behaviour.

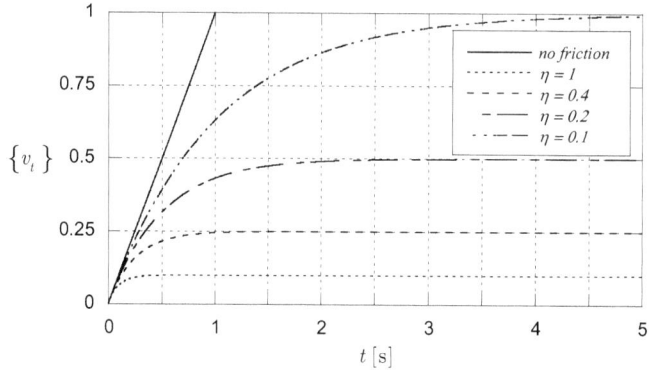

Fig. 4-27: Terminal velocity due to boundary friction for constant tangential acceleration $a_t = 1$ m/s², $\left|\vec{F}_n(\delta_w)\right| = 50$ N, $\mu_v = 1.0033$

4.4.2.3 Wall Bounded Flow

For wall bounded flow that can be described by a logarithmic velocity profile as introduced in section 3.4.2.3, the specification of the terminal velocity of a fluid particle closest to the wall based on the characteristics of the flow may be desired. According to equation (3.50), the velocity $v_t = \overline{u}(\delta_{w0})$ at an equilibrium distance δ_{w0} from the wall for a given roughness k_s can be obtained by

$$v_t(d_{w0}, u_*) = 2.5 \ln\left(\frac{29.7\delta_{w0}}{k_s}\right) u_* \;. \tag{4.140}$$

Furthermore, consider uniform open channel flow with flow depth h_f and slope S_b both together corresponding to roughness k_s. According to equation (2.3), the bottom shear stress is $\tau_b = h_f \rho_f g S_b$ and $u_* = (\tau_b/\rho_f)^{1/2}$. With regard to the weight of the water column above the particle and in accordance with equation (3.38), the tangential shear force and the normal force at the channel bottom are given by

$$\left|\vec{F}_t(\delta_{weq})\right| = \tau_b A = h_f \rho_f g S_b \Delta s^{n-1} \;, \tag{4.141}$$

$$\left|\vec{F}_n(\delta_{weq})\right| = mg \;, \tag{4.142}$$

where Δs is the initial particle spacing, n is the dimension of the problem and $m = h \rho_f \Delta s^{n-1}$.

4.4 Fluid-Structure Interaction

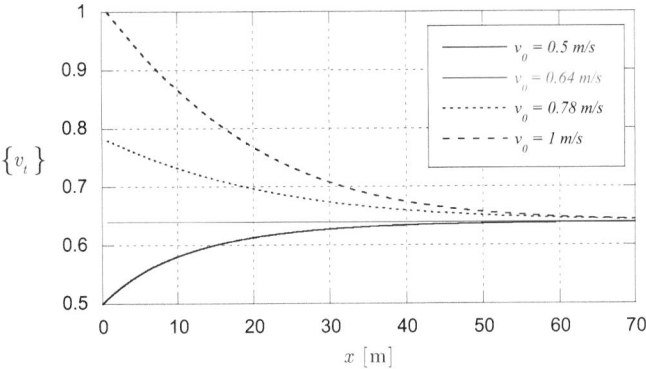

Fig. 4-28: Path of a fluid particle under the influence of a frictional boundary for different initial velocity v_0.

For the present work, uniform flow conditions are of major importance. To reduce the extent of the computational domain, they may be obtained fastest by setting appropriate initial conditions. Nevertheless, in most cases some distance from the channel inlet is needed for the flow to be fully developed. This fact can be illustrated based on the above considerations with focus on a single fluid particle at an equilibrium distance δ_{weq} from the boundary and with initial velocity v_0 as depicted in Fig. 4-28. There, the terminal velocity is assumed to be $v_{end} = 0.64$ m/s, $\eta = 1$, $h_f = 1$ m and $S_b = 0.003$.

As can be seen, the terminal velocity is reached at $x \approx 70$ m in all four cases. However, if the initial velocity is smaller than the desired terminal velocity, a good approximation will be obtained at $x \approx 35$ m already, whereas the approach from a larger initial velocity still has sizeable deviation at that position. Thus, from this point of view, initial conditions based on a velocity smaller than, or equal to the terminal friction velocity at the boundary are preferable.

4.4.3 Damping

For a fluid particle interacting with a structure, damping can be included in a similar way as for the interaction of two spheres as introduced and illustrated in sections 4.3.1.2 and 4.3.2.1, respectively. However, the case of a fluid particle moving close to a boundary with a relative tangential velocity v_t that is moved out of its equilibrium position deserves some attention. Such a case is depicted in Fig. 4-29 in terms of the ratio of an instantaneous dissipative force $F_D = |\vec{F}_D|$ to the maximum dissipative force $F_{D,\max} = |\vec{F}_{D,\max}|$ as well as the ratio of the tangential to terminal velocity v_{end}, whereas the dissipative force consists of friction and damping. Thus, the particle is moved out of its initial equilibrium position by a velocity v_n in the normal direction towards the boundary, i.e. perpendicular to v_t. The size of v_n has been chosen sufficiently large so that the particle loses contact with the boundary during rebound, which is indicated by piecewise vanishing values of $F_D/F_{D,\max}$. As expected, with application of damping the oscillations will disappear after some time and the terminal velocity will be reached. However, rather surprising is the fact that also without damping (and only with partial contact) a similar state is reached where the particle

4 Numerical Methods

oscillates around the terminal velocity. This may be explained by an integral consideration of the dissipative forces, a topic not studied further in this work.

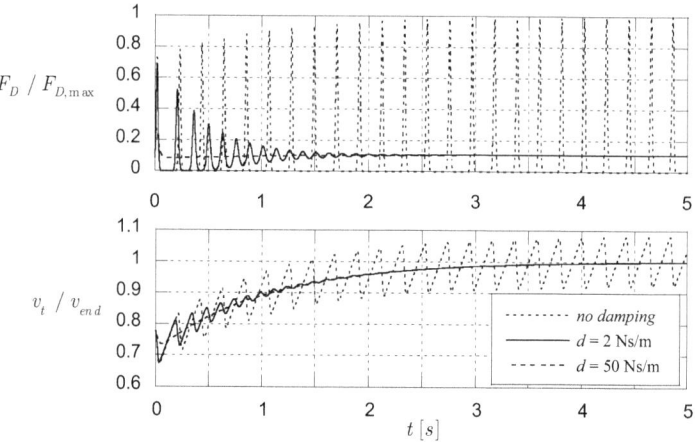

Fig. 4-29: Fluid particle oscillating close to a boundary with and without damping. The ratio of the instantaneous dissipative force to the total dissipative force as well as the ratio of the tangential to terminal velocity is shown.

4.4.4 Time Integration and Solution Algorithm

For the combination of DEM and SPH, the use of different time integration algorithms can lead to an asynchronism, resulting in an unstable simulation. Hence, the same integration scheme with identical parameters is preferred. Therefore, the use of the PC-leapfrog integrator of SPH (as introduced in section 4.2.7.1) for both methods is suggested. The solution algorithm for fluid-structure simulations is a fusion of that for SPH and the algorithm for DEM, where the original steps remain the same and the time-step size is determined by a combination of conditions (4.46) and (4.99).

5 MODEL CALIBRATION AND VALIDATION

5.1 Introduction

In the present chapter the combination of the two numerical methods, SPH and DEM, is applied to study fundamental fluid-structure problems and to evaluate the numerical methods in terms of their capabilities and limits. The applied models are validated by comparing the results of the test cases with reference solutions obtained by physical or empirical relations from the literature. If necessary, the relevant model parameters will be varied in terms of a model calibration until the result of the numerical experiment is in reasonable agreement with reference data. The applications comprise a hydrostatic buoyancy experiment, the settling of a rigid body in a tank filled with water and the simulation of shear flow in an open channel. For the experiments, the size of the fluid particles in terms of their initial particle spacing Δs is chosen several times smaller than the diameter d_s of the corresponding DEM particle. In the present work, this modelling approach where fluid particles are smaller than the rigid body, say $\Delta s \leq d_s/3$, is termed High Resolution Force Model (HRFM). However, SPH is a continuum scale particle method and thus the fluid forces acting on the rigid body remain an averaged quantity.

Tab. 5-1: Main model parameters used for model calibration and validation.

parameter	symbol	value/type	units
sphere radius	r_s	0.015	m
smoothing length	h		m
artificial viscosity coeff.	α	0.01	-
artificial viscosity coeff.	β	0	-
exponent in equ. of state	γ_p	7	-
XPH coefficient	ε_X	0.5	-
artificial stress coeff.	ε_s	0	-
kernel function	W_{ab}	Gaussian	$1/m^\sigma$

The main model parameters which are used for the numerical experiments studied in this chapter are listed in Tab. 5-1; exceptions occur when specific experiments are considered. The initial particle spacing Δs and other parameters vary according to the configuration. The size of the smoothing length is chosen to be $h = 1.5\Delta s$, which corresponds to 29 and 123 neighbouring

5 Model Calibration and Validation

particles in two and three dimensions, respectively. Since the accuracy of SPH depends on the relation between the number of neighbours and the smoothing of local quantities, this is a good choice but also has its computational cost.

5.2 Buoyancy

The effect of buoyancy has been introduced in section 3.4.1, where also the importance of correct modelling of the buoyancy force is pointed out. Improper representation of the buoyancy force may lead to an incorrect weight of the submerged body. Thus, the resistance of the body against acting fluid dynamic forces may be misleading as well. Therefore, the model behaviour for different particle resolutions in two and three dimensional space is studied as well as relevant model parameters are identified in this section.

5.2.1 Configurations

The numerical buoyancy experiments are carried out in a small tank filled with an initially quiescent fluid, i.e. particles are at rest. The dimensions of the water body are: length $l_f = 0.2$ m and height $h_f = 0.1$ m for the two-dimensional (2D) discretisation and $h_f = l_f = w_f = 0.1$ m in the three dimensional (3D) case. Thus, as a conservative estimate the sound velocity is $c_s = 10\sqrt{2gh_f} \approx 14$ m/s, see section 4.2.3.2. As submerged particle, a sphere with three degrees of freedom and radius $r_s = 0.015$ m is used. Three model configurations have to be distinguished: for case A the sphere is located in the middle of the tank at height $z_s = 0.5 h_f$, for cases B and C the sphere sits on top of fixed spheres arranged in a close packing as depicted in Fig. 5-2 and Fig. 5-3.

Tab. 5-2: Initial particle spacing used for buoyancy experiments and resulting number of fluid particles including boundary particles. The second column indicates the number of fluid particles per sphere diameter.

Δs [m]	$2 r_s / \Delta s + 1$ [-]	number of particles 2D	number of particles 3D	average Δt [s]
0.01	4	266	2456	2.10E-04
0.005	7	920	13018	1.10E-04
0.0025	13	3376	82116	5.30E-05

For the given cases, experiments with different resolution of fluid particles in terms of the initial particle distance Δs, hereafter referred to as particle resolution, are carried out as listed in Tab. 5-2. In addition, the ratio of the number of fluid particles per sphere diameter is given (column 2); it can be seen as an alternative indicator for the level of discretisation. Furthermore, the average size of the computational time step Δt is listed since it depends on the smoothing length (compare section 4.2.7.2). The initial particle discretisation for the studied cases are illustrated in Fig. 5-1 and Fig. 5-2.

5.2 Buoyancy

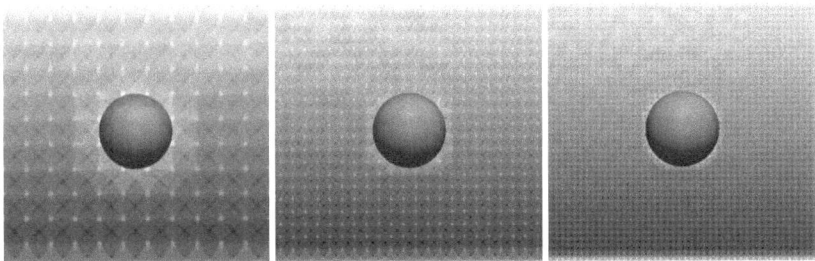

Fig. 5-1: Initially evenly spaced fluid particles in the vicinity of the sphere for case A. The particle spacing is $\triangle s = 0.01$, 0.005 and 0.0025 m from left to right. The shading indicates the hydrostatic pressure, where dark corresponds to larger values than light.

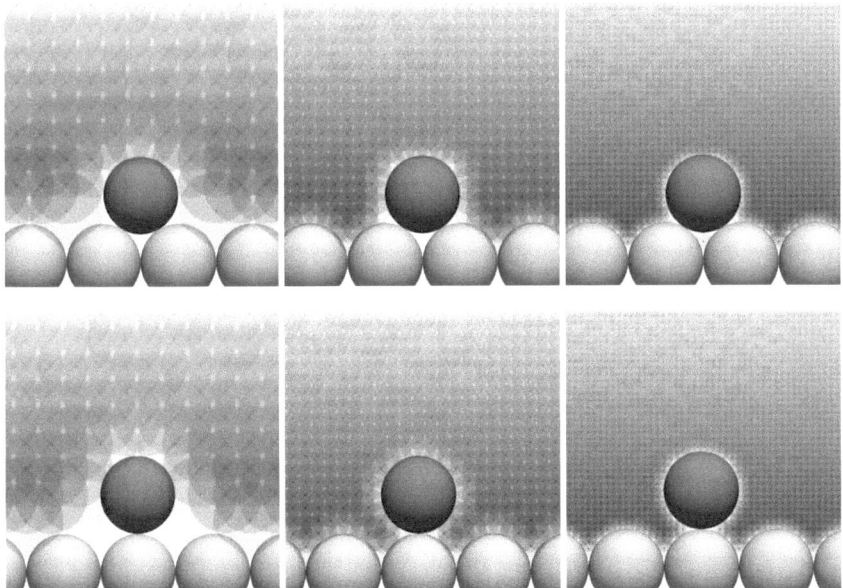

Fig. 5-2: Experimental setup and initially evenly spaced fluid particles in the vicinity of the sphere for case B (upper row) and case C (lower row). The shading indicates hydrostatic pressure, where dark corresponds to larger values than light.

5 Model Calibration and Validation

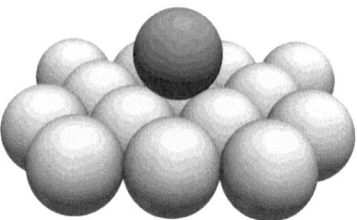

Fig. 5-3: Experimental setup for the three dimensional variant of case B.

The influence of the buoyancy force is studied by varying the density of the sphere, and the resultant submerged weight of the sphere is measured by a kind of load cell connected to the sphere. Actually, the load cell is modelled as a fixed special particle that interacts only with the sphere and not with fluid particles. This special particle overlaps with the sphere, and its initial position corresponds to a penalty force that is equal to the submerged weight. For their interaction, the linear force law is applied with stiffness c_{lc} corresponding to the initial overlap δ_{lc0} and the submerged weight $F_{g*} = |\vec{F}_{g*}|$, i.e. $c_{lc} = F_{g*}/\delta_{lc0}$.

Tab. 5-3: Summary of buoyancy experiments: for case A the density of the sphere is varied to study ascent and descent motion of the sphere. The investigations for cases B and C are limited to experiments with the heaviest sphere.

density of sphere [kg/m³]	case	dim
500	A	2
700	A	2
900	A	2
2800	A, B	2, 3
2800	C	2

For case A, the density of the sphere is varied from 500 to 2800 kg/m³ to study ascent and descent behaviour in 2D. For the heaviest sphere that has the density of granite, cases A and B are also studied in 3D. Situation C is only investigated in two dimensions. The configurations are summarised in Tab. 5-3. Besides the varying arrangement of the sphere of interest in the given cases, the simulation time and progress also differ. For case A, the sphere is free to move from the beginning of the simulation to the end of the measuring period at time 2 s. Alternatively, for cases B and C the sphere is initially fixed and released after half of the simulation time of 4 s. This is so done because in case A the fluid particles are already more or less in place at the initial time compared to the cases B and C where the fluid particles close to the boundary of the solid sphere will rearrange first, trying to fill the gaps between the spheres.

The interaction of the fluid particles with the sphere is modelled by a MLJ potential, since the latter is the preferred force law for fluid-structure interaction. The potential is restricted to repulsive forces and the distance to the sphere surface where the penalty force is zero is set equal to the smoothing length, i.e. $\delta_{w0} = h$, which corresponds to an active penalty force as soon as interaction takes place (compare condition (4.115)). The stiffness of the potential is obtained by evaluating a slightly modified form of equation (8.28) in appendix A, namely

$$k = F\left(\frac{h+r}{h\left(1-\psi_{eq}\right)}\right)^4, \qquad (5.1)$$

where $\psi_{eq} = \delta_{weq}/h$ and the amount of the force is equal to the median pressure acting on the sphere, i.e. at the middle of the sphere $F = \left(h_f - z_s\right)\rho_f g \triangle s^{\sigma-1}$, where σ is the dimension of the problem. The parameter ψ_{eq} actually determines the characteristics of the potential and thus the gradient of the repulsive force. As will be shown later, ψ_{eq} is the main calibration parameter for this experiment. For the given case with almost no fluid motion other force laws would also work, but they may not be able to prevent particle penetration with the estimated parameters when it comes to dynamic problems with larger flow velocities.

5.2.2 Boundary and Initial Conditions

Obtaining appropriate initial conditions for the given case of a quiescent fluid in a tank is quite a challenging task. Despite the fact that all fluid particles are initialized according to a hydrostatic pressure distribution as introduced in section 4.2.5, the initial positions of the fluid particles may not correspond exactly to equilibrium conditions, e.g. in the corners of the tank or at locations where boundary conditions change. Furthermore, the properties of particles, such as the density and consequently the velocity, in contact with a boundary may change since the initial values do not comprise a balance of all local variables. Due to the applied Lagrangian approach, this leads to small displacements of initially evenly distributed fluid particles resulting in equilibrium after some time. Based on these considerations, it seems to be obvious that in such a case, the introduction of damping which consequently generates dissipation leads to a different balanced state that may not represent the physics of the system as desired. The small rearrangement of the fluid particles also affects the forces exerted on the sphere. Thus, use of a load cell seems to be a good choice because after decay of initial disturbances an almost constant value of the force can be measured.

The rigid walls of the tank are modelled by different approaches. The side walls are made of two layers of fixed fluid particles to allow for identical initial hydrostatic pressure conditions as for the fluid particles, since use of triangles would require a scaling of the penalty force according to local pressure which is currently not considered. For the bottom of the tank, three different kinds of boundary particles are briefly examined as follows: (i) two layers of fixed fluid particles, (ii) triangles with penalty force according to a MLJ potential and (iii) with penalty force based on a kernel gradient. In the first case, the boundary particles are initialised with similar properties as the values set for adjacent fluid particles. In the latter cases, the parameters of the force laws are obtained by equations (4.125) and (4.132), respectively: for the MLJ potential $\delta_{weq} = \delta_{w0}/2$, $\delta_{w0} = h$ and for the force based on the kernel gradient $q = 1/2$ as well as the mass m_w is set equal to the mass of an adjacent fluid particle. The force to be balanced corresponds to the weight of the water column on top of a boundary particle, i.e. $F = h_f \rho_f g \triangle s^{\sigma-1}$.

5 Model Calibration and Validation

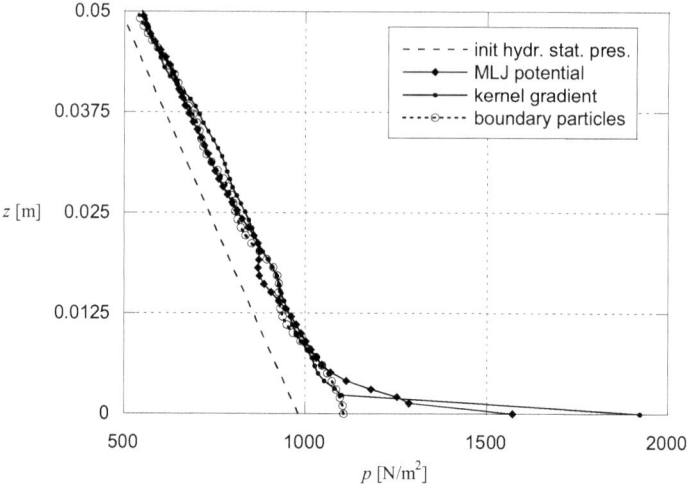

Fig. 5-4: Initial hydrostatic pressure conditions and distributions after a simulation time of 5 s for different boundary conditions at the bottom of the tank.

As noted above, the fluid particles are initialised according to the hydrostatic pressure resulting in the expected distribution (see Fig. 5-4, init hydr. stat. pres.). After starting the simulation, the density of the fluid particles right at the bottom boundary increases; this consequently leads to a rise in pressure until a new local balance is obtained. This effect occurs only at boundaries where the interaction between the particles is treated by a force law. Furthermore, the rise in density is modest compared to that in pressure, which corresponds to the nature of the applied slightly compressible SPH approach using an equation of state. The resulting pressure distributions at a simulation time of 5 s and for the different approaches are depicted in Fig. 5-4. According to the used equation of state, a slight increase of density and corresponding pressure over the whole fluid domain is inevitable. The rise in pressure of fluid particles close to the boundary is only observed for the MLJ potential and the force based on the kernel gradient and not for boundary particles, where only the boundary particles themselves are affected by the pressure increase (not visible in Fig. 5-4).

For the buoyancy experiments, the approach with the penalty force based on kernel gradient is used as bottom boundary condition. Despite the higher pressure at the bottom the disorder of the fluid particles close to the boundary is smaller than for the case with the MLJ potential. Furthermore, an approach with defined boundary force is more generally applicable than the use of boundary particles, especially when it comes to collisions with spheres for example.

5.2.3 Results

Case A : sphere located in the middle of the pool

For the numerical experiments with a configuration according to case A, the parameter ψ_{eq} was varied until the difference ΔF of the exact submerged weight F_{g*} and the force measured by the load cell F_{lc} was within a few per cent, denoted as error in Tab. 5-4.

According to equation (5.1) the force law depends on the parameter ψ_{eq} which actually defines the equilibrium distance between the fluid particles and the sphere by $\delta_{weq} = h\psi_{eq}$. Thus, the parameter ψ_{eq} indirectly controls the amount of displaced fluid and, consequently, the buoyancy force. Furthermore, the mass of the fluid particles is set according to equation (4.37) wherein the term $(\Delta s)^\sigma$ corresponds to a finite area or volume of fluid. Hence, it could be expected that δ_{weq} converges to $\Delta s/2$ for decreasing values of Δs and $\psi_{eq} \to 1/3$ for the present case with $h = 1.5 \Delta s$. This tendency was quite well reproduced by the experiments. However, the determined value for ψ_{eq} varies from experiment to experiment with different density of the sphere. The reason for this is subsequently discussed.

By taking a closer look at the results, it can be seen that the fluid particles arrange in a corona like manner around the sphere as depicted in Fig. 5-5 and Fig. 5-6. This corresponds to the expected behaviour since the pressure acts in the normal direction of the curved surface. However, the resulting pressure distribution around the sphere, i.e. the pressure of the fluid particles in contact with the sphere, is not in agreement with the surrounding fluid particles and is incorrect. Although the final pressure distribution corresponds to an equilibrium state, there are large pressure gradients in the particle corona and fluid particles with relatively small pressure are squeezed. This leads to a slightly different particle arrangement for every experiment resulting in partially inconsistent values for ψ_{eq}.

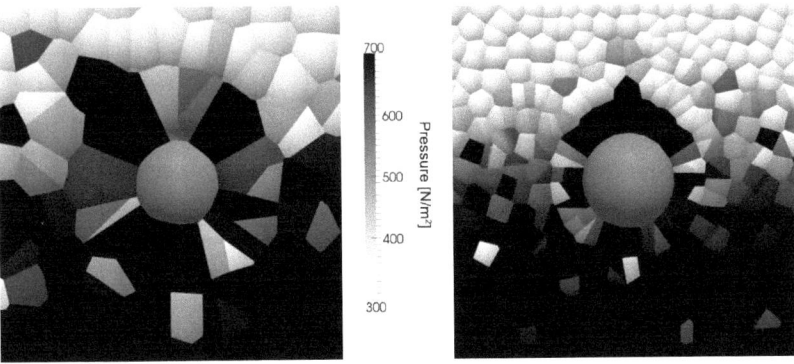

Fig. 5-5: Case A: final arrangement of the fluid particles in the vicinity of the sphere for the two dimensional experiments with $\Delta s = 0.01$ m (left) and $\Delta s = 0.005$ m (right).

5 Model Calibration and Validation

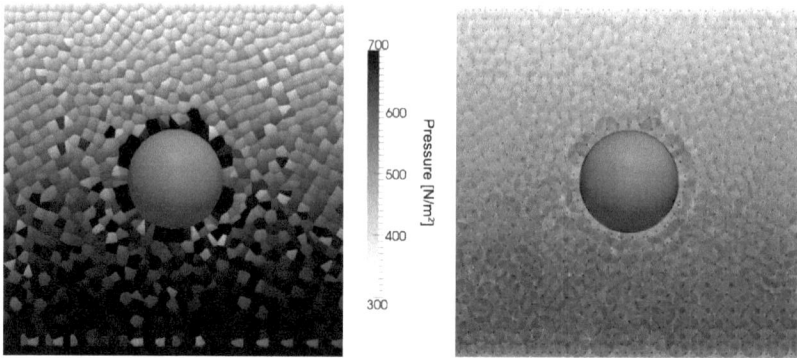

Fig. 5-6: Case A: final arrangement of the fluid particles in the vicinity of the sphere for the two dimensional experiments with $\Delta s = 0.0025$ m. In the picture on the right with transparent particles also the centres of the SPH particles are visible.

Tab. 5-4: Summary of buoyancy experiments for case A with the sphere located in the middle of the tank. The error corresponds to the difference between the exact and measured submerged weight of the sphere, where negative values indicate "lighter than" and positive values "heavier than".

ρ_s [kg/m³]	Δs [m]	ψ_{eq} [-]	F_{g^*} [N]	ΔF [N]	error [%]	comp. time [min]
Case A: 2D experiments						
900	0.01	0.52	0.693	-0.008	1.14	
	0.005	0.4	0.693	0.004	-0.56	
	0.0025	0.43	0.693	0.058	-8.34	
700	0.01	0.518	2.080	-0.024	1.16	
	0.005	0.4	2.080	-0.024	1.17	
	0.0025	0.4	2.080	0.068	-3.28	
500	0.01	0.516	3.467	0.055	-1.60	
	0.005	0.41	3.467	-0.073	2.11	
	0.0025	0.39	3.467	0.033	-0.96	
2800	0.01	0.51	12.482	0.028	-0.22	1.5
	0.005	0.38	12.482	0.090	-0.72	7.5
	0.0025	0.365	12.482	0.154	-1.23	48
Case A: 3D experiments						
2800	0.01	0.49	0.250	-0.001	0.21	48
	0.005	0.405	0.250	0.005	-1.83	650

5.2 Buoyancy

Cases B and C: sphere located on top of fixed spheres

For the configurations B and C the force law between the fluid particles and the sphere is configured in the same way as for case A. Since the situation is similar to the experiments of case A, the above determined parameters ψ_{eq} are also used for the present cases. To allow for the measurement of the submerged weight, no interaction has been specified between the sphere and the fixed spheres at the bottom. Thus, in case B, the sphere sinks a little due to the smaller buoyancy force and the larger submerged weight, respectively. In case C, the sphere is raised a bit due to the fluid particles which squeeze into the gap between the spheres (see Fig. 5-8).

According to the results of case A, the pressure distribution around the sphere is incorrect for the cases B and C; this also holds true for the pressure around the fixed boundary spheres, as expected based on the consideration in section 5.2.2. In Fig. 5-7 and Fig. 5-8, it can be seen that local pressure values of some fluid particles are clearly larger than the expected maximum value at the bottom of 981 N/m². Furthermore, the sphere is initially not completely surrounded by fluid particles as depicted in Fig. 5-2. However, in reality the sphere would be covered by fluid except for the small areas at the contact points. Thus, it could be expected that the exact submerged weight of the sphere is the same as in case A. The simulation results show that for decreasing values of Δs the measured submerged weight converges to the exact value. However, that rate of convergence strongly depends on the particle resolution, i.e. the number of fluid particles located in the spacing between the spheres. This fact confirms the importance of a correct simulation of the buoyancy force.

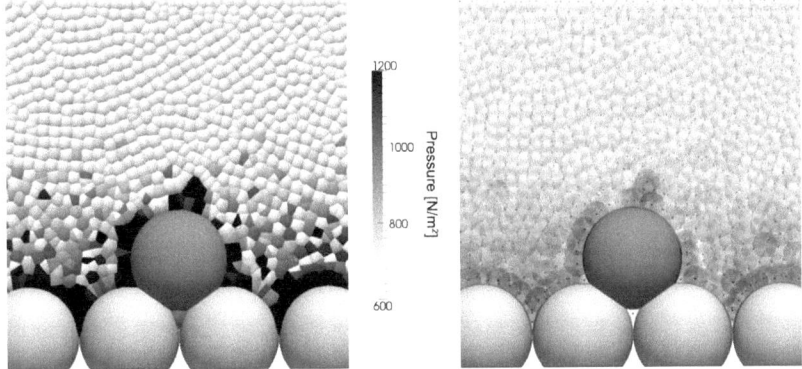

Fig. 5-7: Case B: final arrangement of the fluid particles in the vicinity of the sphere for the two dimensional experiments with $\Delta s = 0.0025$ m. In the picture on the right with transparent particles also the centres of the SPH particles are visible.

5 Model Calibration and Validation

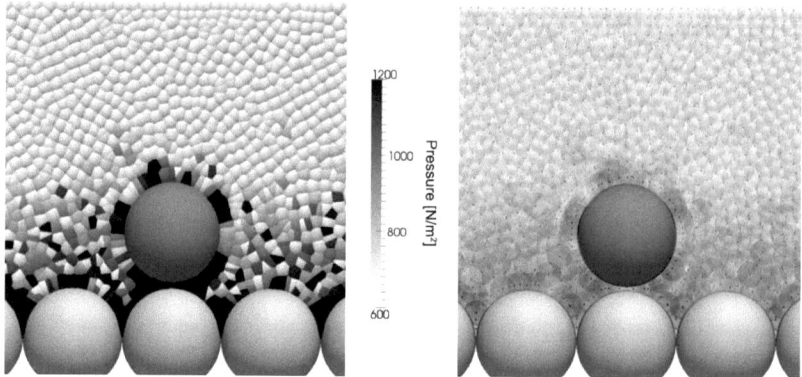

Fig. 5-8: Case C: final arrangement of the fluid particles in the vicinity of the sphere for the two dimensional experiments with $\triangle s = 0.0025$ m. In the picture on the right with transparent particles also the centres of the SPH particles are visible.

To allow for the measurement of the submerged weight, no interaction has been specified between the sphere and the fixed spheres at the bottom. Thus, in case B, the sphere sinks a little after its release at $t = 2$ s due to the smaller buoyancy force and the larger submerged weight, respectively. In case C, the sphere is raised a bit due to the fluid particles which squeeze into the gap between the spheres. This behaviour is illustrated in Fig. 5-9, where the displacement of the sphere for different experiments of cases B and C is shown.

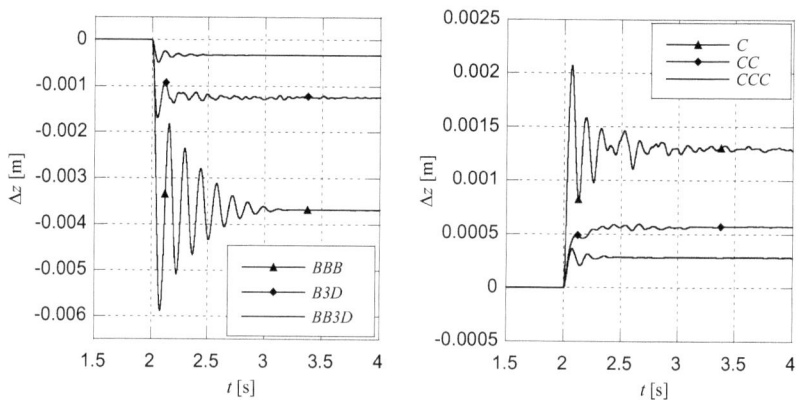

Fig. 5-9: Displacement of the sphere after release at $t=2$ s. The results of case B are shown on the left: 2D experiment with $\triangle s = 0.0025$ m (BBB) and 3D with $\triangle s = 0.01$ m (B3D) and $\triangle s = 0.005$ m (BB3D). On the right the results for case C with initial particle spacing $\triangle s = 0.01$ m (C), $\triangle s = 0.005$ m (CC) and $\triangle s = 0.0025$ m (CCC) are depicted.

Tab. 5-5: Summary of buoyancy experiments for cases B and C with the sphere located on top of fixed spheres. For the ratio F_{lc}/F_{g*} between the measured force by the load cell and the submerged weight of the sphere, values lower than unity indicate "lighter than" and higher values "heavier than".

Δs [m]	F_{g*} [N]	F_{lc} [N]	F_{lc}/F_{g*} [-]	comp. time [min]
Case B: 2D experiments				
0.01	12.482	31.128	2.49	2.5
0.005	12.482	32.853	2.63	14
0.0025	12.482	30.875	2.47	114
Case B: 3D experiments				
0.01	0.250	0.561	2.25	94
0.005	0.250	0.283	1.13	1300
Case C: 2D experiments				
0.01	12.482	6.006	0.48	2.5
0.005	12.482	9.629	0.77	14
0.0025	12.482	11.058	0.89	110

For the buoyancy experiments more than 400 simulations have been carried out. Depending on the configuration, the required computing time is within the range of 1 to 1300 min on a workstation equipped with Intel Xeon X5680@3.3GHz processors. The detailed computing times for some cases are listed in Tab. 5-4 and Tab. 5-5.

5.3 Settling Velocity

A known approach to calibrate and validate the model for fluid-structure interaction is a test case with a sphere falling into a tank of quiescent fluid (compare e.g. Kern and Koumoutsakos (2006)). The calibration comprises the variation of model parameters until the estimated terminal settling velocity can be reproduced by the experiments.

The terminal settling velocity of a sphere in a fluid is reached when the drag force acting on the sphere is balanced by its submerged weight. This can be expressed by addition of equation (3.46) and (3.43) leading to

$$\left|\vec{F}_{dr}\right| + \left|\vec{F}_{g*}\right| = C_D r_s^2 \pi \rho_f \frac{w_s^2}{2} - \frac{4}{3} r_s^3 \pi \left|\Delta\rho\right| g = 0, \quad \Delta\rho = \rho_s - \rho_f. \tag{5.2}$$

By rearranging equation (5.2), the terminal settling velocity for a solid sphere is obtained as

$$w_s = w_{s,sphere} = \sqrt{\frac{8 r_s \left|\Delta\rho\right| g}{3 \rho_f C_D}}. \tag{5.3}$$

The drag coefficient can be estimated by equation (3.47), resulting in $C_D \approx 0.458$ for a sphere with density $\rho_s = 2800$ kg/m^3 and radius $r_s = 0.015$ m. This value of the drag coefficient is an estimate which can be used for the determination of the flow regime, i.e. the estimated terminal settling velocity is $w_s \approx 1.24$ m/s which corresponds to Re $\approx 3.7 \cdot 10^4$. This indicates Newtonian flow for which the drag coefficient has the constant value $C_D = 0.44$. Finally, this results in a terminal settling velocity of $w_{s,sphere} \approx 1.27$ m/s for a sphere with the given properties.

The above considerations are valid for the general three-dimensional case where the body is a sphere. If the situation is reduced to a two dimensional problem, the sphere is replaced by a cylinder with distinct properties. Thus, the drag coefficient and the drag force are also different. Analogous to equation (5.2), the force balance for a cylinder holds

$$\left|\vec{F}_{dr}\right| + \left|\vec{F}_{g*}\right| = C_D d_c \rho_f \frac{w_s^2}{2} - r_c^2 \pi \left|\Delta\rho\right| g = 0, \tag{5.4}$$

where d_c is the diameter and r_c is the radius of the cylinder (see e.g. Jayaweer and Mason (1965)). The corresponding equation for the terminal settling velocity reads

$$w_s = w_{s,cyl} = \sqrt{\frac{r_s \pi \left|\Delta\rho\right| g}{\rho_f C_D}}. \tag{5.5}$$

5.3 Settling Velocity

According to Roshko (1961), the drag coefficient for a cylinder has a constant value $C_D = 1.2$ in the range $2 \cdot 10^4 < \mathrm{Re} < 1 \cdot 10^5$. The resulting terminal settling velocity is $w_{s,cyl} \approx 0.83$ m/s for a cylinder with the given properties and Reynolds number $\mathrm{Re} \approx 2.5 \cdot 10^4$.

For both cases, the cylinder and the sphere, the boundary layer around the body is laminar for the determined Reynolds numbers and the maximum settling velocity. The dominant contribution to the drag force is the pressure drag (compare e.g. Douglas *et al.* (2001)). The minor importance of the friction drag in the considered range of Re is also indicated by the practically constant drag coefficient, where C_D is independent of Re. The influence of the friction drag due to skin friction increases for smaller Reynolds numbers, say $\mathrm{Re} < 10^3$, and also plays a role for the transition of the boundary layer, say around $\mathrm{Re} \approx 10^5$, where the boundary layer becomes turbulent. Thus, for the current experiments the influence of the friction drag is not considered. In the present work, the aspects of the initial particle spacing and the fluid-structure interaction are studied.

5.3.1 Configurations

For the settling velocity experiments, a tank filled with initially quiescent water is considered. The dimensions of the water body are length $l_f = 0.15$ m and height $h_f = 0.3$ m in the case of 2D discretisation and in addition width $w_f = 0.15$ m for 3D. Thus, as a conservative estimate the sound velocity is $c_s = 10\sqrt{2gh_f} \approx 24$ m/s. The radius of the cylinder and the sphere is $r_c = r_s = 0.015$ m and the particle-to-tank width ratio is $\lambda_w = 2r_c/l_f = 0.2$. To account for the effect of wall interference on the settling velocity of the body, the unaffected terminal velocity w_s is reduced to $\hat{w}_s$ according to the empirical relation

$$\frac{\hat{w}_s}{w_s} = 1 - \lambda_w^{1.5} = \gamma_w \qquad (5.6)$$

(see e.g. DiFelice (1996)). For $\gamma_w = 0.91$, the resulting corrected values for the terminal settling velocity of the cylinder and the sphere are $\hat{w}_{s,cyl} = 0.76$ m/s and $\hat{w}_{s,sphere} = 1.16$ m/s, respectively.

Some tests were carried out in advance of the experiments. The effect of the wall interference was tested for case AA by varying the length of the tank, i.e. $l_f = 0.2$ m, $\gamma_w = 0.94$ and $l_f = 0.3$ m $\gamma_w = 0.97$, and the results are in good agreement with the empirical approach above. Different heights of the tank were studied as well, showing no significant effect on the terminal settling velocity. Of course, the height of the tank is increased for the cases with larger settling velocity to prevent premature termination of the experiment due to contact of the body with the tank bottom.

To study the influence of the spatial discretisation, two and three dimensional experiments with different resolution of fluid particles in terms of the initial particle distance Δs are carried out as listed in Tab. 5-6. Similar to the buoyancy experiments, the interaction of the fluid particles with the sphere is modelled by a MLJ potential. Thus, the stiffness of the potential is obtained by equation (5.1), where ψ_{eq} is chosen corresponding to the initial particle distance and the dimensionality according to the parameters obtained by the calibration of the buoyancy experiments (see Tab. 5-4). Since there will be no hydrostatic pressure distribution around the settling body, the reference

5 Model Calibration and Validation

pressure is not known a priori. Hence, the dynamic pressure is taken as reference and the amount of the reference force is obtained by $F = 0.5 \rho_f w_s^2 \Delta s^{\sigma-1}$.

Furthermore, the role of the artificial viscosity and the force law is investigated. The first is studied for four different cases (A, AA, AAA and A3D) by choosing the coefficient $\alpha = 0.0$ (instead of the standard value 0.01), which means that there is no damping of the motion of the fluid particles at all. For the latter, experiments based on case AA with a scaled force law as depicted in Fig. 5-10 are carried out. Compared to the standard configuration of the force law where the influence distance from the wall is equal to the smoothing length, $d_{w0} = h$, the course of the force law is scaled by choosing d_{w0} equal to $2h$, $3h$ and $0.5h$. For $d_{w0} = 3h$, the MLJ potential behaves almost as the linear force law (compare Fig. 5-10).

Tab. 5-6: Initial particle spacing used for settling velocity experiments and resulting number of fluid particles including boundary particles.

case	dim	Δs [m]	number of particles	average Δt [s]
A	2	0.01	620	1.20E-04
AA	2	0.005	2135	6.20E-05
AAA	2	0.0025	7865	3.10E-05
AAAA	2	0.00125	45125	1.50E-05
A3D	3	0.01	12400	1.20E-04
AA3D	3	0.005	74725	6.20E-05

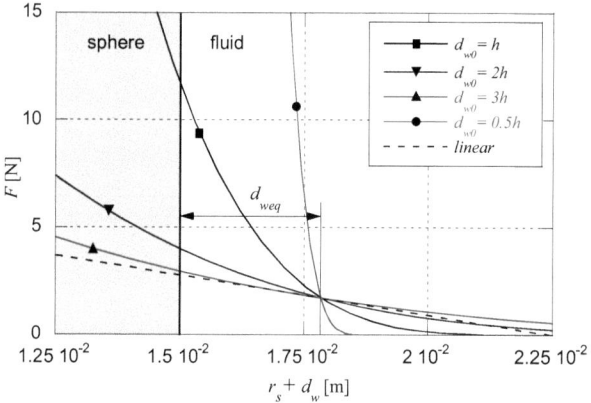

Fig. 5-10: Scaling of the MLJ force law by varying d_{w0}, the distance from the sphere surface where the penalty force is zero. The maximum of the abscissa corresponds to the threshold for interaction of a fluid particle with a sphere, $r_s + h$, according to condition (4.115).

5.3.2 Boundary and Initial Conditions

Similar boundary conditions as those used for the tank of the buoyancy experiments are used for the present experiments. Thus, the walls are made of fixed boundary particles and the penalty force based on kernel gradient is used as bottom boundary condition. The fluid particles are initialised according to a hydrostatic pressure distribution.

The sphere is initially located at height $h_f + r_s$, right above the water surface. To allow for calming of the fluid particles and to minimize the influence of initial perturbations (as discussed in section 5.2.2), the sphere is released after a simulation time of 2 s.

5.3.3 Results

The experiments are evaluated by averaging the settling velocity over time. For a certain time interval Δt, where the settling velocity is approximately constant, the distance covered by the sphere (difference in height Δh_f) is determined; this leads to the observed terminal velocity $\overline{w}_s = \Delta h_f / \Delta t$.

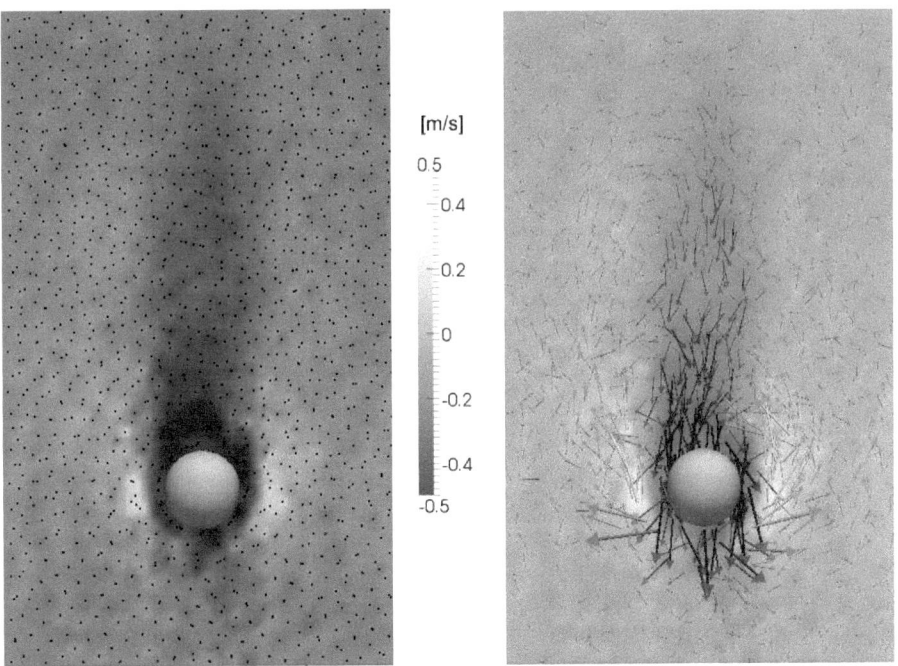

Fig. 5-11: Case AA: contour plot of vertical velocity (left), where points indicate SPH-particle centres, and velocity vectors (right), where the shading indicates magnitude.

5 Model Calibration and Validation

Initial particle spacing and dimensionality

With increasing number of particles, i.e. smaller initial particle spacing, larger terminal settling velocities $\bar{w}_s$ are observed. The measured settling velocity for the coarsest two dimensional particle resolution (case A) is $\bar{w}_s = 0.262$ m/s which increases up to $\bar{w}_s = 0.733$ m/s for the finest resolution (case AAAA) studied in this work.

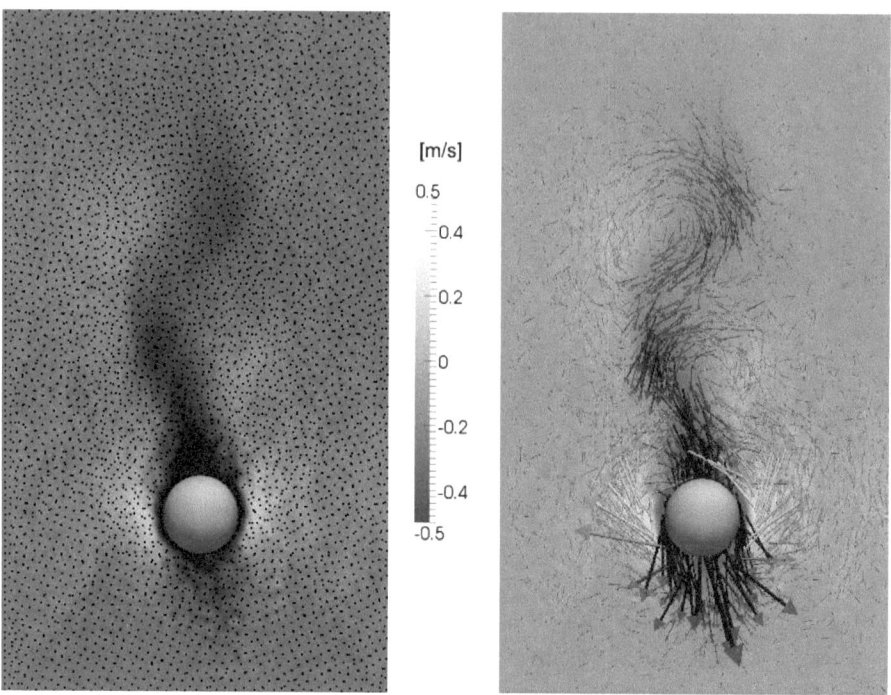

Fig. 5-12: *Case AAA: contour plot of vertical velocity (left), where points indicate SPH-particle centres, and velocity vectors (right), where the shading indicates magnitude.*

For increasing particle resolution, the measured terminal velocity approaches the intended value of $\hat{w}_{s,cyl} = 0.76$ m/s, which indicates convergence of the applied methods. Similar behaviour is rudimentary observed for the three dimensional cases, where $\hat{w}_{s,sphere} = 1.16$ m/s. The results are summarized in Tab. 5-8, p. 122. Furthermore, the flow around the sphere can be reasonably reproduced already for the coarser resolutions; moreover, with smaller initial particle spacing the features of the flow become more detailed, as expected. This is shown by Fig. 5-11 and Fig. 5-12, where the vertical component of the velocity is depicted as contour plot (including particle positions) and the velocity vectors are plotted for selected cases AA and AAA. The topic of how particle data can be interpolated onto a regular grid, to obtain e.g. contour plots, is briefly discussed in section 5.4.3. It has to be mentioned, that besides the different particle resolution, the pictures for the two cases also show distinct flow situations since the velocity of the sphere varies. Despite the

5.3 Settling Velocity

reliable results of the flow field, spurious numerical oscillations in the pressure field are observed. This corresponds to results obtained by other researchers (e.g. Colagrossi *et al.* (2010)), since the standard SPH is known to be noisy.

As already observed in connection with the buoyancy experiments, the fluid particles in the vicinity of the sphere arrange in a corona-like manner. These fluid particles have approximately the same velocity as the sphere for a moment before they move past the body and become part of the wake.

This behaviour is also shown quite plainly by Fig. 5-13, where the vertical velocity for case A3D is illustrated by iso-surfaces. In the vicinity of the sphere, the fluid particles move downwards with the sphere, indicated by the blue drop-like iso-surfaces. These are followed in outward direction of the sphere by a gap of almost zero velocity (no colour) and a belt-like region, where the fluid moves upwards (red).

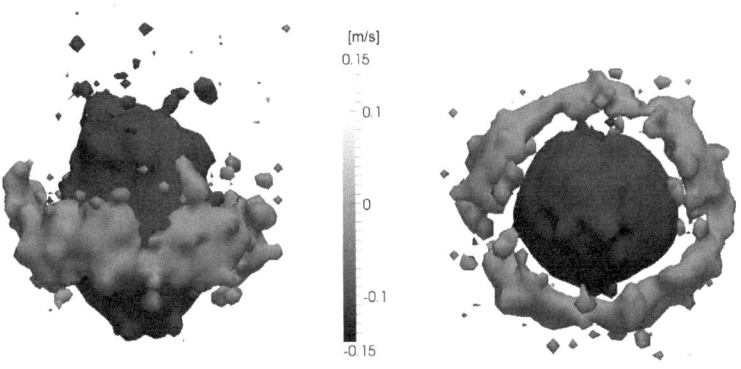

Fig. 5-13: Case A3D: Iso-surface plots of vertical velocity, side view (left) and top view (right).

Based on the above observations, it may be considered that the fluid particles which interact and move with the body downwards are forming together some kind of "meta-particle"[7] with distinct properties. Consider the radius of the "meta-particle" to be $r_m = r_s + \Delta s$. For the two dimensional case, where the meta-particle is a circular disc, this leads to an averaged density of

$$\rho_m = \frac{r_s^2 \rho_s + \left(r_m^2 - r_s^2\right)\rho_f}{r_m^2} \ . \tag{5.7}$$

The corresponding density difference is $\Delta \rho_m = \rho_s - \rho_m$. The results for the meta-particles are summarised in Tab. 5-7. The terminal settling velocity $\hat{w}_{sm}$ for different sizes of the "meta-particle" is obtained by equation (5.5), where the drag coefficient is $C_D = 1.2$ for the determined settling velocities (compare $\text{Re}(r_m, \hat{w}_{sm})$ in Tab. 5-7) and the wall interference is considered by appropriate

---

[7] This larger 'particle' is reminiscent of an added mass.

5 Model Calibration and Validation

values for λ_w. As expected, the calculated terminal velocities $\hat{w}_{sm}$ of the "meta-particle" become larger with decreasing radius r_m and increasing density ρ_m, showing the same trend as for the experiments. However, the range of the terminal velocity is not reproduced and for the case AAAA, the calculated terminal velocity $\hat{w}_{sm}$ is even smaller than the measured velocity $\overline{w}_{sm}$. This indicates that there is actually no such effect due to a "meta-particle" and the dependency of the observed settling velocity on the particle spacing must have another reason.

Tab. 5-7:.Calculation of terminal velocity for different sizes of "meta-particles".

case	r_m	ρ_m	γ_w	$\hat{w}_{sm}$	$\mathrm{Re}(r_m, \hat{w}_{sm})$	$\overline{w}_{sm}$
	[m]	[kg/m³]	[-]	[m/s]	[-]	[m/s]
A	0.0250	1648	0.808	0.521	25787	0.262
AA	0.0200	2013	0.862	0.622	24628	0.475
AAA	0.0175	2322	0.887	0.684	23705	0.622
AAAA	0.0162	2534	0.899	0.719	23148	0.733

A more reasonable explanation can be derived from a more detailed view of the momentum exchange between the sphere and the fluid particles. To this end, consider the collision of a larger sphere with smaller rigid particles which are semi-circularly arranged as depicted in Fig. 5-14. The sphere has the same properties as used for the settling velocity experiments ($m_s = 1.98$ kg for the two dimensional cases) and an initial velocity of $v_1 = v_2 = 0.5$ m/s at time t = 0 s. The smaller particles have the density of water and are initially at rest. Two examples of the momentum exchange are studied: case 1 with five small spheres of radius 0.01 m and mass 0.31 kg (Fig. 5-14 a) and b)), and case 2 with seven even smaller spheres with radius 0.005 m and mass 0.08 kg (Fig. 5-14 c) and d)). As can be seen from Fig. 5-14, in the first case the velocity of the large sphere after collision, v_1, is distinctly smaller than the velocity v_2 in the second case. This fact is also depicted in Fig. 5-15, where the change of the velocity of the sphere is plotted. In addition, the course of the conservative force acting on the sphere is depicted. The maximum of the conservative force occurring and the duration of the impact is smaller in case 2 than in case 1, i.e. also the difference in momentum is smaller: $\int F_{c2} dt < \int F_{c1} dt$. This is because also the sum of the mass of the small particles is smaller in the second case, resulting in a reduced momentum exchange compared to the first case. Furthermore, the velocity of the small sphere in direction of the collision is distinctly larger in the second case than in the first case, which can be seen as another indication for the faster progress of the large sphere in the second case.

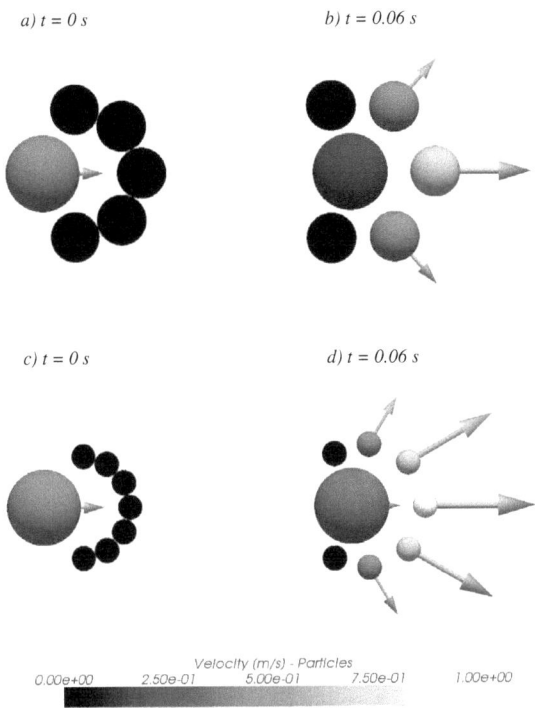

Fig. 5-14: Examples of momentum exchange between a large sphere and smaller rigid particles. In figure b), the velocity vector of the large sphere is not visible on this scale.

These findings can now be transferred to the settling-velocity experiment and reasonably summarized in the following way: for larger initial particle spacing, where also the mass of the fluid particles is larger, the settling of the sphere is hindered more than in the cases with smaller and lighter fluid particles. Furthermore, the instantaneous momentum exchange is also smaller for smaller fluid particles. This fact can also be confirmed by a Fourier transform of the time varying conservative force of the settling sphere, which shows for smaller fluid particles an increasing dominant frequency while the magnitude decreases. This can also be seen from the course of the conservative forces depicted in Fig. 5-15.

5 Model Calibration and Validation

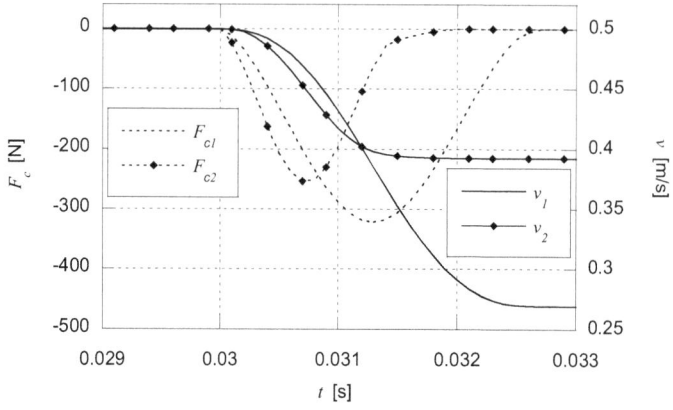

Fig. 5-15: Change of the velocity of the sphere due to collision with smaller particles and course of the conservative forces. The indices indicate cases 1 (five small particles) and 2 (seven small particles).

Influence of artificial viscosity

Application of artificial viscosity in the momentum equation is a possibility to introduce damping in SPH (compare section 4.2.4.1) and thus to increase the stability of the simulation. However, depending on the choice of the parameters, it may also cause the fluid to be more viscous than desired which influences the motion of the settling body. For the present experiments, the standard values of the artificial viscosity coefficients for free surface flows are used, $\alpha = 0.01$ and $\beta = 0.0$. This conservative choice leads to a basic stabilization of the simulation and seems to have no negative effect on the fluid flow. Nevertheless, for the cases A, AA, AAA and A3D, simulations are carried out with no artificial viscosity at all, i.e. $\alpha = \beta = 0.0$. As expected, this leads to a slower decay of the initial perturbations and to less calming of the fluid before the release of the sphere. The desired effect of a reduced drag is only observed for cases A and AA, where for the latter the measured terminal velocity $\overline{w}_s$ increases from 0.475 m/s to 0.536 m/s. For case AAA the opposite effect is observed and the resulting velocity is smaller than with application of artificial viscosity. This may be due to the still distinct perturbations of the flow field when the sphere is released.

The artificial viscosity may play a role concerning the terminal settling velocity. However, it seems only reasonable to study a possible effect by a parameter variation when the terminal velocity resulting from an experiment with standard configuration is close to the desired value. However, this would require an even higher particle resolution than case AAAA and thus more time or computational power, which would exceed the scope of the present work.

Scaling of force law

The influence of the scaling of the force law on the terminal settling velocity is studied for case AA. In this case, the force law is scaled by varying d_{w0}, the wall distance of the point where the repulsive force becomes zero. Besides the standard configuration with $d_{w0} = h$, three different setups are studied: $d_{w0} = 2h$ and $d_{w0} = 3h$, resulting in a boundary condition with a smaller maximum repulsive force and slower increase for decreasing particle distance, and $d_{w0} = 0.5h$, which has the opposite effect (compare Fig. 5-10). Notice that the amount of the reference force at wall distance d_{weq} does not change for the investigated cases.

The reference force which also defines the characteristic of the MLJ potential is assumed to be equal to the maximum hydrodynamic pressure, as introduced in section 5.3.1. However, the total pressure acting on the sphere is actually larger, because the a priori unknown ambient pressure is not considered. For the case with the standard configuration, $d_{w0} = h$, the difference of the reference force and the total effective pressure is less important, since the deviation is compensated by the force law in terms of a slightly smaller wall distance than the supposed equilibrium distance. For this configuration, the fluid particles are arranged around the sphere in the expected manner and the majority of the particles do not penetrate the sphere surface, as depicted in Fig. 5-16.

In the two cases where $d_{w0} > h$, the increase of the repulsive force is slower than for the standard configuration and the deviation mentioned before becomes more apparent. The fluid particles are now able to penetrate the surface of the sphere. Thus, the equilibrium distance is smaller than the radius of the sphere and corresponds to a repulsive force which is equal to the local effective pressure. This has the effect, that the buoyancy force of the sphere is reduced and the terminal settling velocity is larger than for the standard configuration, i.e. the measured terminal velocity $\bar{w}_s$ increases from 0.475 m/s ($d_{w0} = h$) to 0.593 m/s ($d_{w0} = 3h$).

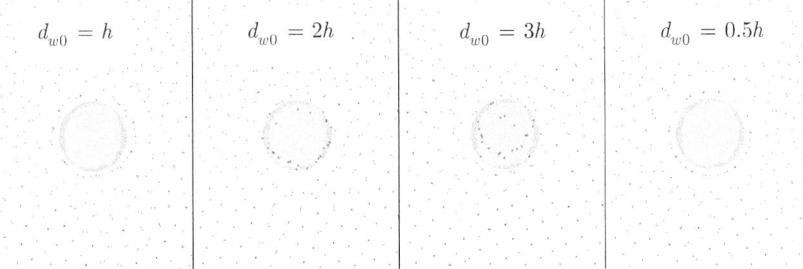

Fig. 5-16: Positions of fluid particles for different scaling of the MLJ force law. The pictures show the sphere at a position of approximately $0.5h_f$.

The increase of the stiffness of the MLJ potential, as in the case where $d_{w0} = 0.5h$, has only a marginal effect on the settling velocity. The measured terminal velocity $\bar{w}_s$ slightly increases from 0.475 m/s ($d_{w0} = h$) to 0.489 m/s ($d_{w0} = 0.5h$) (see Tab. 5-8). A possible explanation for the larger terminal velocity may be that the rebound of the fluid particles interacting with the sphere is faster due to the larger penalty force at the boundary. However, the larger stiffness is also a potential source of numerical instability, since very large pressure gradients between fluid particles of the corona around the sphere are observed.

5 Model Calibration and Validation

Tab. 5-8: *Summary of settling velocity experiments, where α is the artificial viscosity coefficient, d_{w0} is the distance from the surface of the sphere where the penalty force is zero, and $\overline{w}_s$ is the observed settling velocity.*

case	α	d_{w0}	$\overline{w}_s$	comp. time
	[-]	[m]	[m/s]	[min]
initial particle spacing and dimensionality				
A	0.01	h	0.262	12
AA	0.01	h	0.475	83
AAA	0.01	h	0.622	537
AAAA	0.01	h	0.733	*1660
A3D	0.01	h	0.167	1497
AA3D	0.01	h	**0.16	27626
artificial viscosity				
A	0.00	h	0.359	
AA	0.00	h	0.536	
AAA	0.00	h	0.545	
A3D	0.00	h	0.229	
AA3D	0.00	h	0.294	
scaling of force law				
AA	0.01	$2h$	0.506	
AA	0.01	$3h$	0.593	
AA	0.01	$0.5h$	0.489	

*) without initial calming; simulation time = 0.6 s
**) no constant settling velocity achieved after simulation time of 4.5 s

For the settling-velocity experiments more than 170 simulations have been carried out. Depending on the configuration, the required computing time is within the range of 10 to 27600 min on a workstation equipped with Intel Xeon X5680@3.3GHz processors. The detailed computing times for some cases are listed in Tab. 5-8.

Impacting of sphere on free surface

For the sake of completeness and to exemplify one of the strengths of the applied methods, namely SPH, for case AAA the immersion of the sphere after its release is shown in Fig. 5-17. The sphere is released from height $h_f + r_s$, i.e. the velocity at the initial impact is zero; this leads to a smooth plunge of the sphere. For the presented experiment, neither surface tension nor the correction for free surface flows (XSPH) is considered. Despite the rather coarse particle resolution, the deformation of the initially plane water surface and the splashing are reproduced in a characteristic manner for a sphere with hydrophobic surface as observed by other researchers (see e.g. Do-Quang and Amberg (2010)). The hydrophobic behaviour (see e.g. Duez *et al.* (2007) leads to a cavity behind the impacting sphere. This may be expected, since the present configuration of the applied force law between the sphere and the fluid particles only considers repulsive forces.

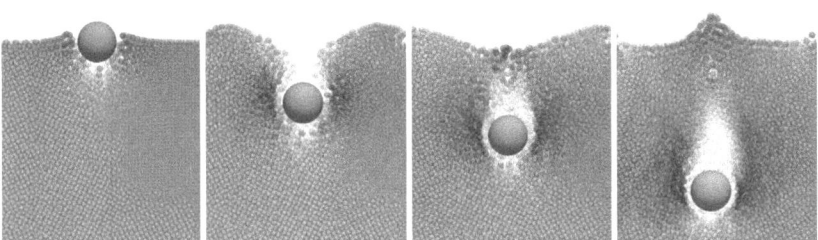

Fig. 5-17: Case AAA: Impacting of a sphere on a free surface after its release at t=3 s. From left to right: t=3.56, 3.65, 3.7 and 3.78 s. Shading indicates the vertical velocity: lighter negative, darker positive with respect to the z-direction.

5.4 Open Channel Flow

A common approach for the experimental investigation of sediment transport is to provide steady uniform flow conditions in a straight laboratory flume to obtain a constant and calculable situation for the experiment. For such a situation and with respect to the cross-sectional flow velocity, it is considered that the flow conditions neither change with time (steady) nor with position (uniform) along the channel. For an open channel with a rough bottom boundary, steady uniform flow conditions can be obtained by slight inclination of the channel bed to balance the energy loss due to friction by the reduction of the potential energy. Thus, the slope of the water surface and the energy grad line will become equal to the slope of the channel bed S_b. The resulting flow depth is called normal or uniform depth h_n.

With regard to a cross-section of the channel, the corresponding distribution of the velocity component parallel to the channel bed can be described by the logarithmic law for turbulent channel flow given by equation (3.50) in general. However, since a small flow depth h_f compared to the roughness height d_s, i.e. a small relative flow depth h_f/d_s, is preferred for the present experiments (compare next section 5.4.1) a modified form of the log law has to be applied. Bezzola (2002) presents a modified log law for such situations, which reads for a very wide channel where side wall effects are negligible

$$\frac{\overline{u}(z)}{u_*} = c_R \left(\frac{1}{\kappa} \ln \frac{z}{y_R} + 8.48 \right). \tag{5.8}$$

The corresponding average flow velocity is given by

$$\overline{u}_m = c_R \left(2.5 \ln \left(\frac{10.9 h_f}{y_R} \right) \right) \sqrt{g h_f S_b}. \tag{5.9}$$

For $h_f/y_R > 2$, the coefficient c_R in equations (5.8) and (5.9) is $c_R = \sqrt{1 - y_R/h_f}$, where the height of the roughness sublayer for a plane bed with uniform grains is $y_R \approx 1.0 d_s$.

5.4.1 Configurations

To enable a reasonable parameter study for the simulation of open channel flow, the configuration has to match with the available computational and temporal resources. Since the simulation time and the total number of particles are the primary factors which determine the computing time, the particle resolution and the geometric variables defining the computational domain, i.e. the flow depth and the channel length, have to be chosen appropriately. Furthermore, the experiments are reduced to a vertical two-dimensional configuration and two different particle resolutions are studied, namely $\Delta s = 0.01$ m (case A) and $\Delta s = 0.005$ m (case AA).

For the open channel flow experiments, the slope of the channel bed is $S_b = 0.0035$ and the proposed flow depth is $h_f = 0.09$ m. Thus, as a conservative estimate the sound velocity is $c_s = 10\sqrt{2gh_f} \approx 13$ m/s. For the standard grain or sphere diameter, $d_s = 0.03$ m, the relative flow depth is $h_f/d_s = 3$, the height of the roughness sublayer is $y_R \approx 1.0 d_s = 0.03$ m and the parameter $c_R \approx 0.82$. According to equation (5.9), this leads to an average flow velocity of $\overline{u}_m = 0.4$ m/s.

Based on preliminary test experiments, the length of the channel is chosen to be 1.5 m. It is observed that this length is large enough to enable the development of the characteristic vertical velocity distribution and to minimize the boundary influence. In combination with a special inflow condition (see section 5.4.2), approximate uniform flow conditions are obtained after a flow distance of 8 to 10 h_f and the influence of the outflow boundary is 2 to 4 h_f, both depending on the particle resolution. As observed in the test experiments, steady flow conditions are obtained after a simulation time of approximately 20 s. Thus, the total simulation time is 30 s, chosen so as to allow for time averaging of the results.

5.4.2 Boundary and Initial Conditions

Three different boundary conditions are used for the channel flow experiments. At the upper end of the channel, an inflow boundary condition according to section 4.2.6.1 is defined, where particles are continuously moved into the domain with uniform velocity $\overline{u}_m$. The channel bottom at the inflow boundary has the form of a small ramp as depicted in Fig. 5-18, which has the effect that the flow is locally accelerated in the direction away from the boundary. This is very important to prevent numerical instabilities in the area of the inflow boundary, which may be caused by local squeezing of fluid particles. The squeezing originates from the deceleration of the flow due to bottom friction, and the backwater effects due to the weir at the outflow boundary before the flow reaches a steady state.

The weir at the outflow boundary serves as pressure boundary and minimises the extent of the backwater curve. Right after the weir, at the end of the channel, a particle sink is located, which eliminates any fluid particles. To obtain a continuous outflow of the particles, a weir with an inclination of 45 degrees is used (see Fig. 5-18). An estimate for the weir height can be obtained based on Poleni's formula,

$$h_w = h_f - \left(\frac{q_{we}}{C_w}\right)^{2/3}, \qquad (5.10)$$

where $C_w = (2/3)\mu_w\sqrt{2g}$. The discharge coefficient μ_w for a sharp crested weir with 45 degrees streamwise inclination is $\mu_w \approx 0.84$ (see e.g. Bollrich (1996)); thus the coefficient $C_w \approx 2.48$. The specific discharge for the present experiments is $q_{we} = \overline{u}_m h_f = 0.036$ m²/s. Evaluating equation (5.10) for the given values leads to an estimated weir height of $h_w \approx 0.03$ m.

The bottom of the channel consists of triangles. The interaction between the triangles and the fluid particles is modelled by a MLJ potential. The stiffness of the potential is obtained from equation (4.125). For the present experiments, the distance from the wall where the repulsive force is zero is $d_{w0} = h$ and the equilibrium distance is chosen as $d_{weq} = 0.5h$ ($d_{weq} = 0.0075$ m for case A and $d_{weq} = 0.00375$ m for case AA). The amount of the reference force corresponds to the proposed maximum hydrostatic pressure at the channel bottom, i.e. $F = \rho_f h_f g \Delta s$.

5 Model Calibration and Validation

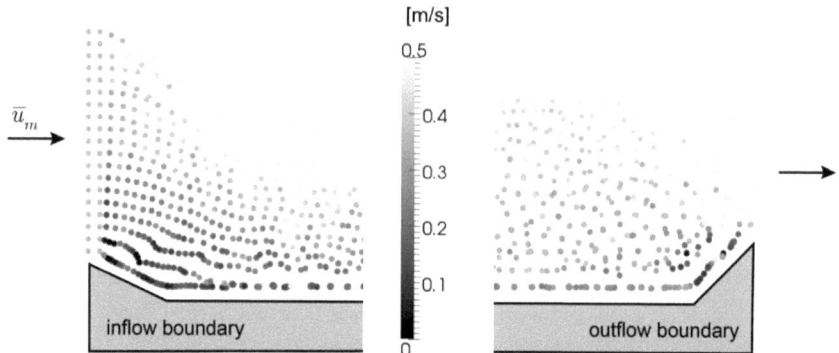

Fig. 5-18: *Geometry of the channel bed at the in- and outflow boundaries. At the inflow boundary, a ramp is used (left) to accelerate the flow and at the outflow boundary, a weir is used to achieve an appropriate flow depth at the boundary.*

The channel roughness is modelled by the friction law according to equation (4.133). An estimate for the friction coefficient μ_v can be obtained based on the considerations described in chapter 4.4.2.2. The velocity u_w close to the channel bottom at distance $z = d_{weq}$ is obtained by evaluation of equation (5.8), where the shear velocity is $u_* = \sqrt{gh_f S_b}$, i.e. $u_w \approx 0.23$ m/s for case A and $u_w \approx 0.15$ m/s for case AA. Thus, by solving equation (4.138) for a normal force $|F_n| = F$, velocity $v_t = u_w$ and $\eta = 0.1$, leads to an estimated friction coefficient of $\mu_v \approx 0.154$ for case A and $\mu_v \approx 0.235$ for case AA.

Due to the special channel geometries at the in- and outflow boundaries, a proper initialisation of the channel by fluid particles with initial velocity is quite complicated. Thus, an empty channel was considered as initial condition.

5.4.3 Results

For the two particle resolutions and the given geometry of the channel, the weir height h_w, the coefficient μ_v controlling the boundary friction at the channel bottom and the coefficient α for the artificial viscosity are varied until the desired velocity profile is achieved. In the scope of the open channel flow experiments, more than 70 simulations were carried out with an average computing time ranging from 360 min (case A) to 2370 min (case AA) for a simulation time of 30 s. Based on the findings obtained from the experiments, the different effects of the varied model parameters can be categorised as depicted in Fig. 5-19. The velocity of the fluid particles close to the channel bottom is controlled by the parameters for the boundary friction. The upper half of the velocity profile significantly responds to changes of the weir height and the intermediate part of the profile is sensitive to variations of the artificial-viscosity coefficient.

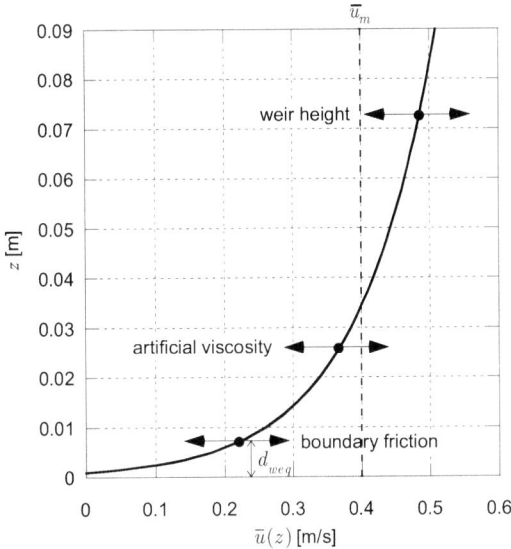

Fig. 5-19: Effect of the different model parameters which are varied to obtain the desired velocity distribution of open channel flow.

As pointed out by Jang et al. (2009), the evaluation of SPH data is not as straightforward as for grid based data. To generate profile or contour plots the properties of fluid particles, such as the velocity, have to be interpolated onto a regular set of points by scattered data approximation. A common approach for that is inverse distance weighted interpolation. In this work, Shepard's method is preferred (see e.g. Holger (2009)). Another approach is to use SPH particles as described by Jang et al. (2010), which is only useful if the particles are well arranged and the particle density is sufficiently large. For disordered particles the interpolation error increases; this limits the significance of the results.

For both cases, the particle data is interpolated onto a vertical profile located at a distance of 1.2 m from the inflow boundary. Since the fluid particles at the boundaries, i.e. the channel bottom and the free surface, only have neighbouring particles on one side, the interpolation is poor at these locations. Close to the free surface, the interpolation error is small because of the almost uniform velocity distribution in this region. However, the interpolation data of the fluid particle right at the free surface is discarded, which leads to an offset of approximately $\Delta s/2$ at the top of the profile. At the channel bottom where the velocity gradients are large, the interpolation error leads to misleading profile data. Thus, only profile data above a certain offset from the bottom boundary are considered. Since the fluid particles which move closest to the channel bottom already have an offset of d_{weq}, the total offset at the bottom boundary is approximately $d_{weq} + \Delta s/2$.

5 Model Calibration and Validation

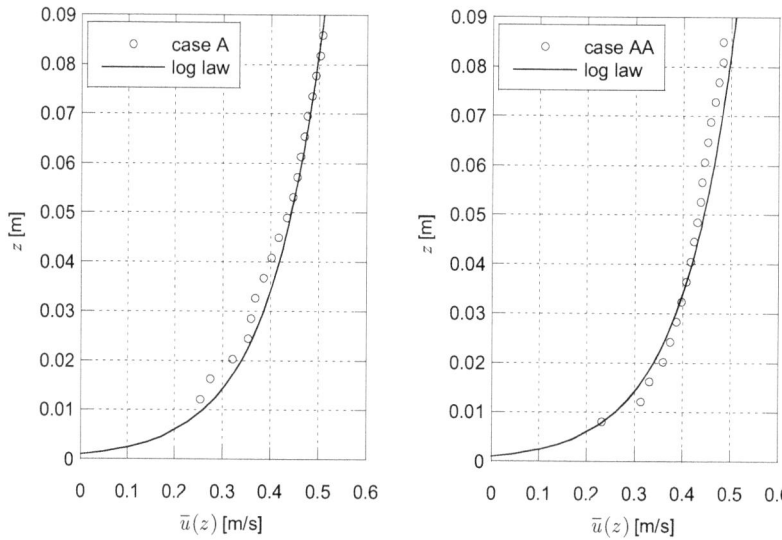

Fig. 5-20: Calculated velocity profiles of the open channel flow experiment, case A and case AA. The simulation results are marked by circles which correspond to interpolation points and not to fluid particles.

Furthermore, the profile data are temporally averaged to obtain a smoother velocity profile. Since the flow becomes steady after a simulation time of 20 s, the interpolated velocities are averaged from 20 to 30 s. The resulting velocity profiles for cases A and AA are depicted in Fig. 5-20 and are in good agreement with the theoretical velocity profile according to equation (5.8). The results show that in case AA the region of the velocity profile with large velocity gradients can be better reproduced than in case A with coarser particle resolution. The corresponding model parameters obtained by calibration are for case A, $h_w = 0.0275$ m, $\mu_v = 0.16$, $\alpha = 0.005$ and for case AA, $h_w = 0.025$ m, $\mu_v = 0.21$, $\alpha = 0.03$. Comparison of the calibrated model parameters with the above estimates for the weir height and the friction coefficients confirm the usefulness of the approaches used for the estimation.

5.5 Discussion

The chosen experiments have proven to be a reliable concept to evaluate the numerical methods which are used to model the fluid-structure interaction. For the applied high resolution force model (HRFM), several model parameters which are decisive to obtain reliable simulation results were identified. Besides the parameters defining the interaction between SPH and DEM particles, the particle resolution plays a major role, especially when it comes to dynamic problems. Also numerical parameters such as the coefficient for artificial viscosity are important for the stability of the simulation as well as for boundary layer flows. Furthermore, the expected convergence of the SPH method is demonstrated and approved by the results of the experiments.

Nevertheless, some drawbacks of the applied methods have to be pointed out. As shown by the validation experiments, the computational cost is already quite high for two-dimensional simulations with a moderate number of particles. This limits the scope of parameter studies, and three-dimensional investigations become very time-consuming. Hence, in the present work the influence of the boundary friction and the artificial viscosity on the terminal settling velocity was for example not further investigated. To overcome these limitations, parallelisation of the software is necessary to allow for the use of high performance computing infrastructure. Another common approach to essentially reduce the computing time and to study local flow phenomena would be the implementation of periodic boundary conditions. However, the implementation of these approaches is challenging for particle methods because the exchange of information across contiguous boundaries is quite sophisticated (see e.g. Fleissner (2010)).

Furthermore, the significance of the pressure increase at the boundaries between fluid particles and rigid walls (fixed and movable) is not clear. For the present work, this particularly applies to the settling velocity experiments where the pressure drag results from the pressure distribution around the sinking body. The problem may have two contributions, namely the spurious numerical oscillations in the pressure field and the fluid-structure boundary condition itself. Some recent contributions by other researchers provide approaches which may be useful to overcome this shortcoming. The pressure oscillations are typical for standard SPH and are due to the weakly compressible approach. Colagrossi and Landrini (2003) suggest the filtering of the density with a moving-least-square integral interpolation, which leads to a smooth and reliable pressure distribution for free surface flows (compare Colagrossi *et al.* (2010)). Their corrected SPH approach has been further improved by Molteni and Colagrossi (2009). A similar approach is presented by Bonet and Kulasegaram (2002), and Becker *et al.* (2009) point out its advantage over the standard SPH in combination with a special boundary treatment for fluid-structure interaction. An alternative approach may be the application of truly incompressible SPH (iSPH), as outlined in chapter 4.2.1.2, where the incompressibility of the fluid is ensured by additional constraints which lead to a correct pressure distribution. An approach how to capture the local pressure for a point of a rigid movable boundary based on the adjacent fluid particles is presented e.g. by Maruzewski *et al.* (2010).

Boundary conditions for fluid-structure interaction are still an open and challenging task for movable arbitrary shaped bodies as discussed in 4.2.6.3. For piece-wise straight fixed boundaries, the use of boundary particles (dynamic particles) with copied properties of the adjacent fluid properties leads to exact free-slip conditions (see e.g. Colagrossi and Landrini (2003)). However, this approach may lead to undesired behaviour for models where flow separation occurs due to possible attractive effects at the boundary as already mentioned in section 4.2.6.3. A more promising approach, suitable for the recent applications, is presented by Monaghan and Kajtar (2009). For arbitrarily shaped particles, they suggest to use boundary particles which exert forces on the fluid particles. By appropriate choice of these forces and of the spacing of the boundary particles they obtain good results for various experiments with a cylindrical body.

Another inconsistency of the applied methods arises due to the force law used to model the interaction between the fluid particles and the rigid bodies. For the present experiments, the parameters of the force law depend on the particle resolution and on a reference force. The latter corresponds to a specific equilibrium condition which is different for the buoyancy and the settling velocity experiments. As can be seen from the experiments, appropriate boundary forces are crucial to correctly describe either situation. However, for the modelling of sediment transport both situations are appearing at the same time in a simulation. Thus, a boundary force which adapts to the effective conditions would be preferable. According to the convergence of the SPH method, the difference in the parameters may vanish with increasing particle resolution. However, this limits the application of the methods to a certain particle resolution. This problem may be solved by implementation of a dynamic force calculation based on the current flow variables or by application of a different boundary condition as discussed above.

It seems to be obvious that as long as the shortcomings of the current modelling approach are not adequately solved, numerical experiments with focus on detailed local forces are not reasonable. However, these kinds of experiments are necessary to investigate the fundamental physical processes occurring during incipient motion and sediment transport. Possible experiments to validate an improved version of the model would be e.g. the determination of the drag coefficient of a sphere sitting on a boundary of similar spheres as investigated by Coleman (1972) or the incipient motion experiments carried out by Fenton and Abbott (1977).

6 APPLICATION TO SEDIMENT TRANSPORT

6.1 Introduction

To demonstrate the capabilities of the present model for the simulation of sediment transport, a slightly different modelling approach as used in the previous chapter is applied in the following. The applications to sediment transport comprise a two-dimensional simulation of the development of a scour caused by a freefalling water jet and a three-dimensional pier scour experiment. For the present experiments, the size of the fluid particles in terms of their initial particle spacing Δs is intentionally chosen larger than for the model-verification experiments, i.e. $\Delta s = d_s/2$, where d_s is the diameter of the corresponding DEM particle. This modelling approach where fluid particles are larger than the rigid body, say $\Delta s > d_s/3$, is termed Low Resolution Force Model (LRFM). With the LRFM, simulations on a larger scale than with the HRFM are made possible. However, due to the usually larger computational domain and because the sediment layer also consists of particles, no "miracles" concerning the computational costs have to be expected.

Due to the less detailed resolution of the fluid forces acting on a solid particle, the model parameters have to be calibrated to match the desired sediment transport processes; this relates to the spatial as well as the temporal scale. Depending on the complexity of the experiment, the calibration can be quite extensive. For the present experiments only marginal calibration of the model parameters was carried out. Thus, the presented simulation results are rather of qualitative nature and primarily serve for illustration purposes.

6.2 Scour Caused by a Freefalling Water Jet

In the present experiment, the development of a scour due to a freefalling water jet is studied. This kind of scour is typical for a natural waterfall, where at the bottom of the subsequent plunge pool a scour hole develops caused by the impact of the freefalling water. Similar processes can also be observed in the plunging jet pool downstream of hydraulic structures, such as a weir in a river or the spillway of a dam, where the protection of the in-situ river bed is inadequate. Plunge pool scour is an important topic in hydraulic engineering since increasing and uncontrolled scour may lead to a destabilisation of embankments or of the hydraulic structure itself by backcutting erosion.

For the experiment, the specific discharge is 0.04 m²/s and the head drop is 0.16 m. Since the experiment is carried out as a two-dimensional simulation, the sediment consists of circular particles with diameter $d_s = 2r_s = 0.01$ m and density $\rho_s = 2800$ kg/m³. As mentioned in the introduction, the ratio of the sediment diameter to the initial fluid particle spacing is 2, i.e. $\Delta s = 0.005$ m. The density of the fluid is $\rho_f = 1000$ kg/m³ and the sound velocity is $c_s \approx 14$ m/s. With a safty coefficient for the CFL condition of $\alpha_s = 0.3$ and a reference flow depth of $h_f = 0.1$ m, the estimated hydrodynamic time-step size is $\Delta t = \alpha_s h_f/c_s \approx 2 \cdot 10^{-3}$ s and the estimated sediment time-step is $\Delta t = \alpha_s r_s / c_s \approx 1 \cdot 10^{-4}$ s; thus the latter is relevant. The total simulation time is 15 s and the corresponding computing time for the experiment was approximately 1060 min on the same hardware used as for the model-verification experiments.

For the interaction between the sediment particles, Hertz's law is applied. Based on the considerations of section 4.3.7, Young's modulus for the sediment particles is chosen as $E_k = 4 \cdot 10^8$ N/m². This results in a practicable size of the time step and a realistic behaviour of the sediment particles. The effective time-step size is $\Delta t \approx 3 \cdot 10^{-5}$ s, because it is determined based on the external forces. The sediment particles are considered to consist of granite which has a Poisson's ratio of $\nu_k = 0.25$ (see appendix A.1.2). The internal friction properties of the sediment layer are chosen according to appendix A.1.4, i.e. equal coefficients for sticking and slipping friction $\mu_s = \mu_k = 0.7$.

The interaction between the fluid and sediment particles is modelled by an MLJ potential. Due to the applied LRFM approach also the concept for the parameters of the force law is different to that used for the model verification. For the present experiments, the distance from the sediment-particle surface where the repulsive force is zero is $d_{w0} = h - r_s$ and the equilibrium distance is chosen as $d_{weq} = 0.5 d_{w0}$. The stiffness of the potential is 600 N. With this configuration, the sediment particle may behave like a heavy fluid particle when it encounters true fluid particles (notice that this only concerns the fluid-sediment interaction). The parameters for the friction between the fluid and sediment particles are $\mu_v = \eta = 0.1$.

The results of the simulation at selected times are shown in Fig. 6-1. At the beginning of the simulation, the sediment erosion advances quickly due to the unimpeded impact of the water jet on the sediment surface. Already after some seconds a scour hole develops and the water depth at the impact location increases. The resulting plunge pool now alleviates the momentum of the water jet and sediment erosion diminishes. After a simulation time of about 10 s, the extent of the scour hole will barely change. The development of the scour, i.e. the profile of the bed level, is reproduced in a characteristic manner by the numerical model (compare e.g. D'Agostino and Ferro (2004).

As can be seen from Fig. 6-1, the fluid particles are able to enter the sediment layer up to a certain depth, which is similar to seepage. This behaviour is depicted on the left in Fig. 6-2 in detail where the black dots indicate the locations of fluid particles. The fluid particles fill up the voids between the sediment particles. The exerted forces by the fluid particles in the pores may be interpreted as a mix of buoyancy and lift forces.

6.2 Scour Caused by a Freefalling Water Jet

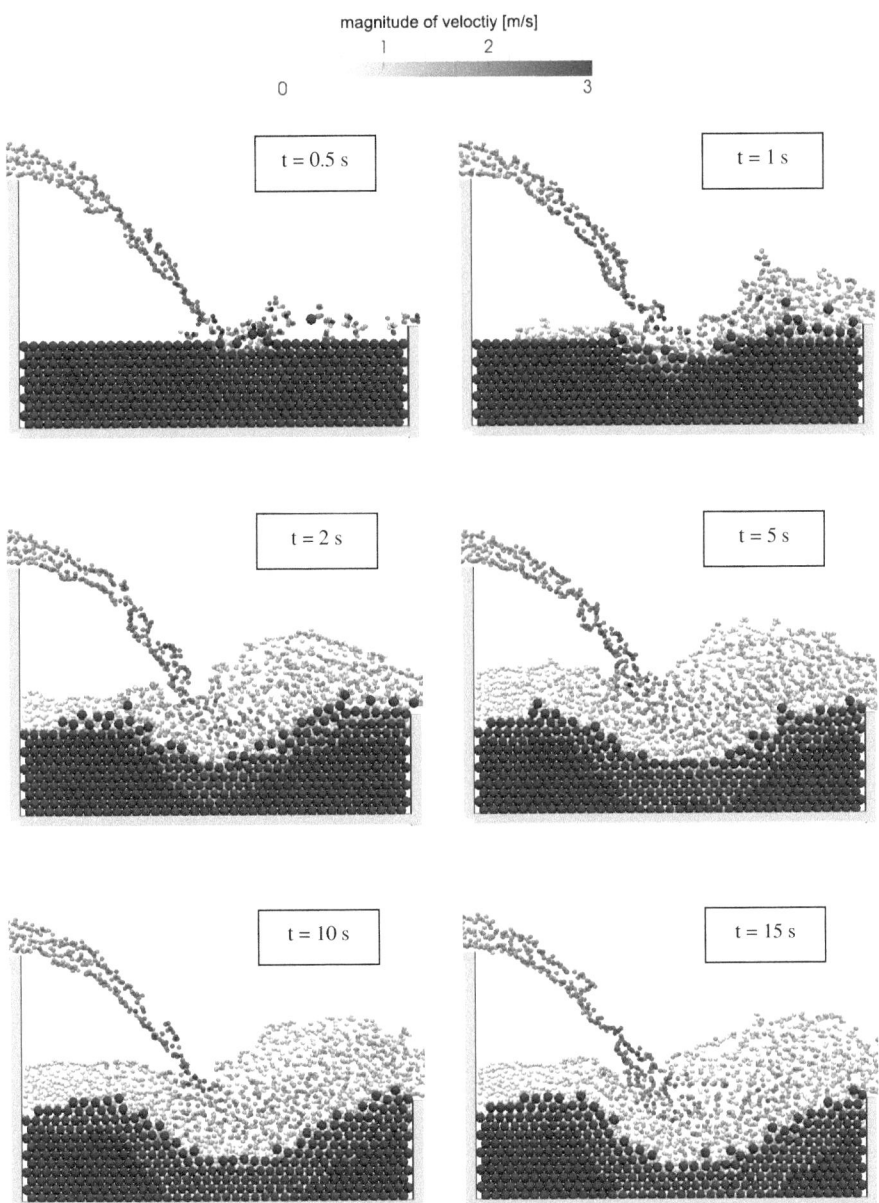

Fig. 6-1: Numerical simulation of a scour caused by a freefalling water jet.

6 Application to Sediment Transport

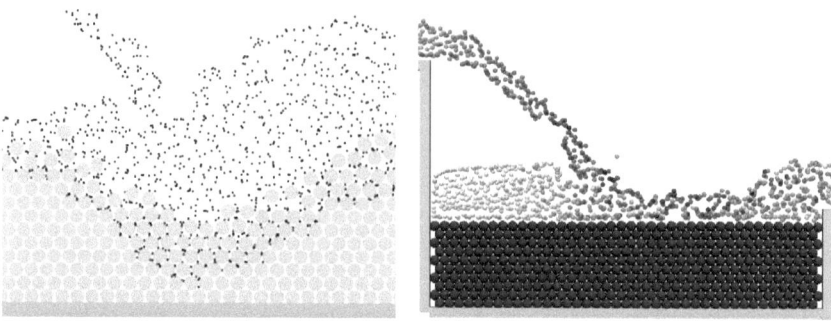

Fig. 6-2: *Influence of interaction law on the erosion process. In the left picture, the result of a simulation with an interaction law based on $d_{w0} = h - r_s$ is depicted (compare Fig. 6-1), where the black dots indicate the locations of the fluid particles. On the right, the simulation result for a different configuration with $d_{w0} = h + r_s$ is shown.*

The major importance of the possibility for fluid particles to enter the sediment layer for the present experiments can be illustrated by varying the parameters of the force law for the fluid-sediment interaction. For this purpose, consider a different configuration: the distance from the sediment-particle surface where the repulsive force is zero is increased to $d_{w0} = h + r_s$ while the stiffness, i.e. the size of the maximum repulsive force, is kept constant. This leads to a repulsive force which already acts at a distance between a fluid particle and the sediment-particle surface which is larger than for the previous configuration with $d_{w0} = h - r_s$. Furthermore, a reduced sediment density of $\rho_s = 1800$ kg/m³ is considered to emphasize the distinct behaviour. The simulations show that the fluid particles are no longer able to move between the sediment particles (compare Fig. 6-2, on the right), which is due to the scaling of the force law. Despite the reduced mass of the sediment, no scour is observed at all.

For the given situation, the basic difference between the present model and a Lagrangian two-fluid-continuum approach is how seepage and buoyancy effects are taken into account. For the first, seepage is controlled by the parameters of the interaction force law between fluid and sediment particles, as discussed above. For the latter, where the sediment and the water are treated as separate fluid phases with different densities, seepage effects have to be considered in the continuum description, e.g. by adding a force term to the momentum equation (see e.g. Bui *et al.* (2007)).

6.3 Clear-Water Scour at Bridge Pier

Another well-known kind of scour is the erosion of sediment observed at bridge piers. Some part of the water which impinges the pier is vertically deflected in the downward direction. This leads to formation of a horseshoe vortex close to the bottom which extends along the pier. Right downstream of the pier the formation of a wake vortex is observed. The deflection of the water at the pier and the resulting vortices lead to an erosion of sediment and the development of a characteristic scour hole around the pier (see e.g. Unger and Hager (2007)). The understanding of the erosion process at bridge piers is important to apply effective protection measures and consequently prevent a possible failure of the structure.

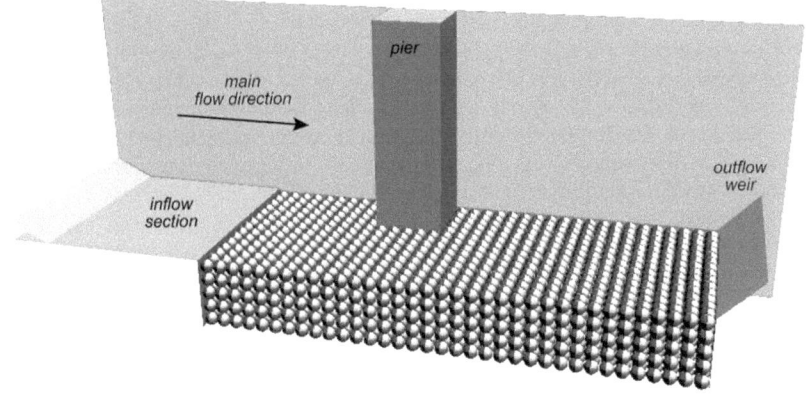

Fig. 6-3: Configuration of the pier-scour experiment.

The channel used for the three-dimensional pier-scour experiment is 0.7 m long and 0.15 m wide. The slope of the channel is 3.5 ‰. The quadratic pier with a side length of 0.05 m is located at a distance of 0.25 m from the inflow boundary. It is placed adjacent to the wall as depicted in Fig. 6-3. The bottom of the channel consists of two parts: a fixed inflow section of 0.15 m length is followed by a movable sediment bed. The sediment consists of five layers of particles which results in a total of 2484 sediment particles. For the three-dimensional simulation, the sediment consists of spheres with diameter $d_s = 0.01$ m and density $\rho_s = 2500$ kg/m³. The ratio of the sediment diameter to the initial fluid particle spacing is two. The initial water height at the inflow boundary is 0.1 m and the inflow velocity is 0.4 m/s, i.e. the discharge is 0.04 m³/s. The initial spacing of the fluid particles is $\Delta s = 0.005$ m and the density of the fluid is $\rho_f = 1000$ kg/m³. Based on the chosen particle resolution the total number of fluid particles during the simulation is approximately 70'000. To obtain appropriate outflow boundary conditions, a weir is placed at the end of the channel which is only effective for fluid particles. The size of the time step is similar to that observed in the jet scour experiments. Furthermore, the simulation starts from an initially dry channel bed as depicted in Fig. 6-3.

Similar to the jet scour experiment, Hertz's law is used to model the interaction between the sediment particles. The corresponding parameters are $E_k = 4 \cdot 10^8$ N/m^2 and $\nu_k = 0.25$. Since the energy of the flow is distinctly smaller than in the previous experiment, the friction between the sediment particles is reduced to trigger a faster erosion process, i.e. $\mu_s = \mu_k = 0.45$. For the interaction between the fluid and sediment particles a MLJ potential is applied. The parameters of the force law are $d_{w0} = h - r_s$ and $d_{weq} = 0.5 d_{w0}$. For the three-dimensional case the stiffness of the potential has to be strongly reduced compared to the previous 2D experiment and is 3 N. The parameters for the friction between the fluid and sediment particles are $\mu_v = \eta = 0.1$. The interaction of fluid particles with the sidewalls is modelled as a MLJ potential without friction (no-slip condition). For the interaction with the pier friction is considered.

The results of the experiment at different times of the simulation are depicted in Fig. 6-4 and Fig. 6-5, where the sidewalls, the inflow section and the outflow weir are omitted to improve visibility. At the beginning of the simulation, the waterfront moves across the domain and the weight of the water causes a small depression of the sediment-bed surface. However, no significant transport of sediment particles takes place at this point. At the pier, some part of the water which impinges the pier is vertically deflected in the downward direction. With increasing flow depth also this effect amplifies. The downward flow exerts larger contact forces on the sediment particles than on those exposed mainly to tangential flow. At the first, this leads to initial transport of some sediment particles and to local erosion in front of the pier. Subsequently, due to the initial erosion the transport of sediment particles is amplified and the erosion extends around the pier and along the channel with time.

Due to numerical instabilities the computation was aborted after a simulation time of 1.6 s, which corresponds to a computing time of 17 days. The origin of the instabilities is similar to that observed at the open channel-flow experiments. The pier causes a backwater effect at the left side of the channel, which leads to the instabilities close to the left part of the inflow boundary. To prevent these instabilities the inflow section would have to be extended to a multiple of its present length, which consequently would lead to a significant increase in computing time. Nevertheless, the simulation results show the initial phase of the erosion process, which is reproduced in a reliable manner and which is comparable to experimental observations (see e.g. Radice *et al.* (2006), Radice *et al.* (2008)).

6.3 Clear-Water Scour at Bridge Pier

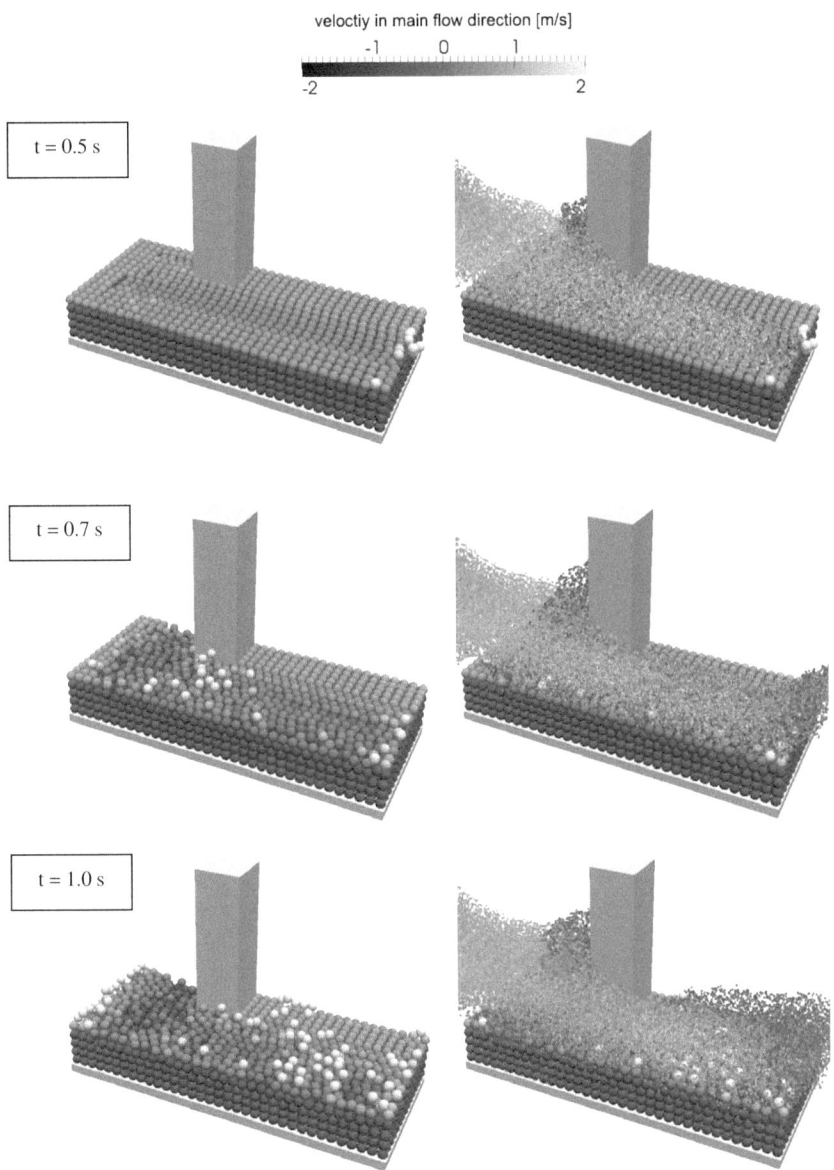

Fig. 6-4: Numerical simulation of a clear-water scour at bridge pier. A lighter colour of the sediment particles indicates "higher above datum". The sidewalls, the inflow section and the outflow weir are omitted to improve visibility.

6 Application to Sediment Transport

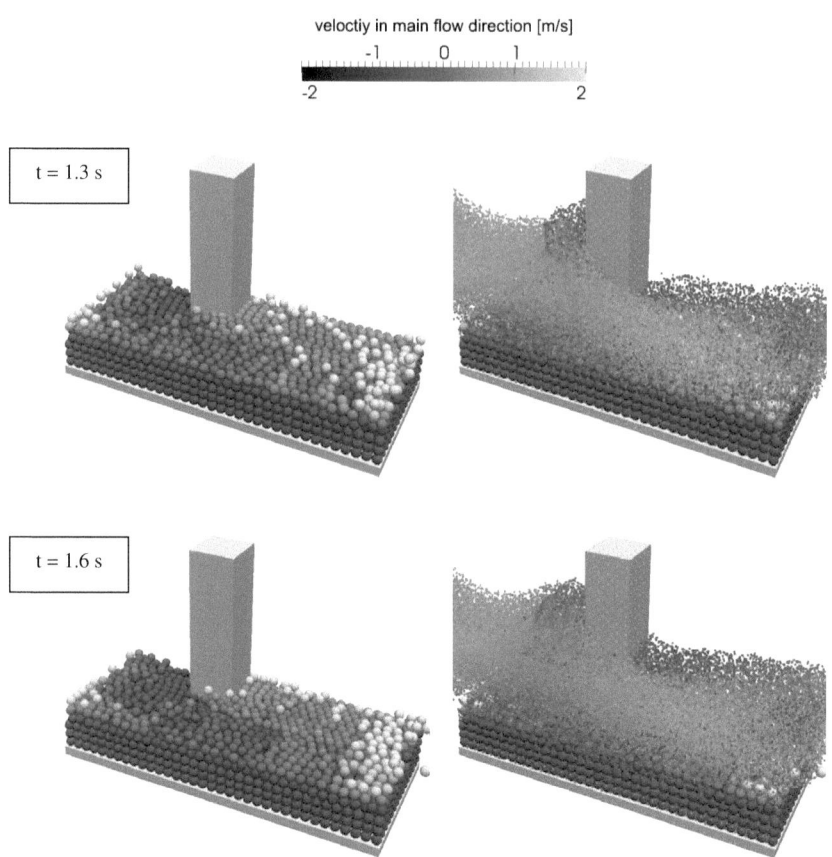

Fig. 6-5: Numerical simulation of clear-water scour at bridge pier. A lighter colour of the sediment particles indicates "higher above datum". The sidewalls, the inflow section and the outflow weir are omitted to improve visibility.

7 CONCLUSION

7.1 Summary

In the present work, a novel numerical model for the simulation of sediment transport with regard to different spatial resolutions is presented. The model consists of the combination of two strictly meshfree Lagrangian methods which allow for the simulation of fluid-structure interaction problems. The interaction between the fluid and the sediment particles and between the sediment particles themselves is modelled by a well-defined force law also accounting for various kinds of friction between the grains. For the present work, the sediment grains are modelled as spherical particles. The applied model is able to reproduce the constitutive behaviour of sediment mixtures and the different transport modes of bed load, such as sliding, rolling and saltating.

The modelling of the fluid is based on a continuum approach which is discretised by the Smoothed Particle Hydrodynamics (SPH) method. The sediment particles are represented by the Discrete Element Method (DEM), where the interactions between the discrete sediment grains are modelled by a force law, which is also able to account for various kinds of friction. A similar approach is applied to the interaction between the fluid and sediment particles. The definition of the interface and the exchange of forces between the fluid and sediment grains are inherent to the applied approaches. Thus, the use of a computational grid or of techniques for the tracking or capturing of the interface is not necessary.

The different force laws relevant for the present subject are presented in detail. To simplify their configuration and to provide appropriate equilibrium conditions, relations derived from the force laws, in terms of the distance of the fluid particle from the solid boundary and a reference force, are provided. These can be used for the determination of relevant model parameters and allow for an expedient configuration of the simulations. Furthermore, a procedure to estimate the material stiffness depending on characteristic simulation quantities and the standard deviation of the penalty force is presented. The reason for this is that the SPH method requires an explicit scheme for time stepping. Thus, to limit the corresponding restriction of the time-step size a certain error may be taken into account, while allowing for a faster progression of the simulation. Moreover, a procedure is presented for the estimation of the friction parameters for wall-bounded shear flows depending on a proposed near wall velocity.

Two basically different approaches to model sediment transport with the proposed method are presented. On the one hand, the application of the combined methods as a High Resolution Force Model (HRFM) is investigated. For the HRFM, the fluid particles are chosen distinctly smaller than the sediment particles to simulate detailed interaction forces. To study the interaction forces on a spherical particle depending on the resolution of the fluid particles a hydrostatic and a dynamic experiment, namely the simulation of buoyancy effects and the determination of the settling

7 Conclusion

velocity, are carried out. The simulations show convergence of the results for increasing particle resolution; they turned out to be a reliable concept to validate the chosen numerical approaches. Furthermore, the importance of the possibility to account for the effect of buoyancy is pointed out. In addition, channel flow experiments are carried out and the relevant model parameters for wall bounded shear flow have been identified. The simulation results show the potential of the HRFM to be used for detailed investigations of bed load processes.

On the other hand, the use of the model in terms of a Low Resolution Force Model (LRFM) is studied. For the LRFM, the fluid particles are chosen of similar size or larger than the sediment particles. This requires a basically different approach for the determination of the interaction-force law parameters. Due to the less detailed resolution of the fluid forces acting on a solid particle, the model parameters have to be calibrated to match the desired sediment transport processes; this concerns the spatial as well as the temporal scale. Furthermore, the solid particles may no longer only represent a single sediment grain, but rather a small volume of sediment or a chunk of soil. Depending on the complexity of the experiment, their calibration may become quite extensive. Hence, the presented simulations are only marginally calibrated. The LRFM was applied to scour caused by a freefalling water jet and to clear-water scour at a bridge pier. The qualitative simulation results are in satisfying agreement with experimental observations and illustrate the use of the applied methods for practical applications.

The methods also have some shortcomings. The force law used for the interaction of the fluid and the sediment grains depend on a reference force, which may not correspond to the actual and local fluid forces in a dynamical simulation. This may affect the accuracy of the results. However, this effect mainly applies to the HRFM, and its influence is expected to diminish for increasing particle resolution. Its role with regard to the forces on a sediment particle embedded or close to the sediment bed has to be investigated in subsequent research. Moreover, boundary conditions between the fluid and solids for SPH are still an open topic. Another problem arises due to the pressure field, which shows spurious oscillations inherent to the weakly compressible SPH method. There are several approaches to overcome this issue, either by the introduction of appropriate filtering or by the implementation of a truly incompressible variant of the SPH method. The mentioned shortcomings are discussed in detail in section 5.5.

As also pointed out by other researchers, the main drawback of the presented model is due to its extensive computational cost for detailed and three-dimensional simulations. For example, the three-dimensional pier-scour experiment with a total particle number of about 80'000 requires about 25 days for a simulation time of 1.5 seconds on a modern high-end processor using one core. The common way to overcome this restriction is to implement parallelisation techniques to be able to use high performance computing infrastructure. However, as far as engineering practice is concerned, the use of the present model in the near future is not realistic, since its appropriate application is still a challenging task and the corresponding computation time requirement may not be affordable.

7.2 Recommendations for Future Research

Subsequent research may include the extension of the present model in two directions. On the one hand, further investigation of problems concerning the numerical approaches on a very detailed level (as with the HRFM) is important. On the other hand, research work with the main focus on engineering applications (as with the LRFM) should be carried out.

With regard to the HRFM, alternative approaches concerning calculation of the pressure field and the boundary conditions may be implemented. The approaches which use filtering in combination with appropriate boundary conditions may be expedient. A viable alternative to this would be the implementation of a truly incompressible variant of the SPH method. Another important topic would be the elimination of numerical instabilities by effective approaches to enable faster and more flexible applications of the model. A model which includes these extensions is expected to serve as a research tool for the investigation of the drag and lift forces acting on a sphere exposed to three-dimensional channel flow.

When it comes to the LRFM, the determination of reasonable force law parameters depending on the resolution and the sediment properties is of basic interest. The main question is: "What are the matching parameters for a certain scale?" However, the LRFM is a promising approach in combination with experimental data and may serve as a reliable simulation tool for many hydraulic and environmental engineering problems.

The successful combination of both tasks could bridge the gap between the two modelling approaches and would open the way to true multi-scale modelling.

Bibliography

Adalsteinsson, D. and Sethian, J. A. (1995) 'A fast level set method for propagating interfaces', *Journal of Computational Physics*, 118 (2), pp.269-277.

Alkhaldi, H., Ergenzinger, C., Fleissner, F. and Eberhard, P. (2008) 'Comparison between two different mesh descriptions used for simulation of sieving processes', *Granular Matter*, 10 (3), pp.223-229.

Andersson, S., Söderberg, A. and Björklund, S. (2007) 'Friction models for sliding dry, boundary and mixed lubricated contacts', *Tribology International*, 40 (4), pp.580-587.

Andrew, B. (2005) *Lagrangian fluid dynamics*. Cambridge University Press, Cambridge.

Apperley, L. W. and Raudkivi, A. J. (1989) 'The entrainment of sediments by the turbulent-flow of water', *Hydrobiologia*, 176, pp.39-49.

Armanini, A. (1995) 'Non-uniform sediment transport: dynamics of the active layer', *Journal of Hydraulic Research*, 33 (5), pp.611-622.

Armanini, A., Capart, H., Fraccarollo, L. and Larcher, M. (2005) 'Rheological stratification in experimental free-surface flows of granular-liquid mixtures', *Journal of Fluid Mechanics*, 532, pp.269-319.

Ashby, M. F. and Jones, D. R. H. (2005) *Engineering materials*. Elsevier/Butterworth-Heinemann.

Ashida, K. and Michiue, M. (1971) 'An Investigation of River Bed Degradation Downstream of a Dam', *14th IAHR Congress*. Paris.

Ataie-Ashtiani, B. and Shobeyri, G. (2008) 'Numerical simulation of landslide impulsive waves by incompressible smoothed particle hydrodynamics', *International Journal for Numerical Methods in Fluids*, 56 (2), pp.209-232.

Bagnold, R. A. (1936) 'The movement of desert sand', *Proceedings of the Royal Society of London Series a-Mathematical and Physical Sciences*, 157 (A892), pp.594-620.

Bagnold, R. A. (1941) *The Physics of Blown Sand and Desert Dunes*. London: Methuen.

Bagnold, R. A. (1956) 'The flow of cohesionless grains in fluids', *Philosophical Transactions of the Royal Society of London Series a-Mathematical and Physical Sciences*, 249 (964), pp.235-297.

Bagnold, R. A. (1973) 'Nature of saltation and of bed-load transport in water', *Proceedings of the Royal Society of London Series a-Mathematical Physical and Engineering Sciences*, 332 (1591), pp.473-504.

Baker, G. R. and Beale, J. T. (2004) 'Vortex blob methods applied to interfacial motion', *Journal of Computational Physics*, 196 (1), pp.233-258.

Balevicius, R., Kacianauskas, R., Mroz, Z. and Sielamowicz, I. (2011) 'Analysis and DEM simulation of granular material flow patterns in hopper models of different shapes', *Advanced Powder Technology*, 22 (2), pp.226-235.

Barry, J. J., Buffington, J. M. and King, J. G. (2007) 'A general power equation for predicting bed load transport rates in gravel bed rivers (vol 40, art no W10401, 2007)', *Water Resources Research*, 43 (8).

Batchelor, G. K. (2005) *An introduction to fluid dynamics*. Cambridge University Press, Cambridge.

Bibliography

Bathurst, J. C., Graf, W. H. and Cao, H. H. (1987) 'Bed Load Discharge Equation for Steep Mountain Rivers ', in *Sediment Transfer in Gravel-Bed Rivers*. John Wiley & Sons, New York, pp.453-477.

Becker, M., Tessendorf, H. and Teschner, M. (2009) 'Direct Forcing for Lagrangian Rigid-Fluid Coupling', *IEEE Trans. Vis. Comput. Graph.*, 15 (3), pp.493-503.

Bell, R. G. and Sutherland, A. J. (1983) 'Non-equilibrium bedload transport by steady flows', *Journal of Hydraulic Engineering-Asce*, 109 (3), pp.351-367.

Benz, W. (1990) 'Smooth Particle Hydrodynamics - A Review', *Numerical Modelling of Nonlinear Stellar Pulsations*, 302, pp.269-288.

Bertrand, F., Leclaire, L. A. and Levecque, G. (2004) 'DEM-based models for the mixing of granular materials', *5th International Symposium on Mixing in Industrial Processes*. Jun 01-04. Seville, SPAIN: pp.2517-2531.

Bettess, R. (1984) 'Initiation of sediment transport in gravel streams', *Proceedings of the Institution of Civil Engineers Part 2-Research and Theory*, 77 (MAR), pp.79-88.

Bezzola, G. R. (2002) Fliesswiderstand und Sohlenstabilität natürlicher Gerinne unter besonderer Berücksichtigung des Einflusses der relativen Überdeckung. *VAW Mitteilung* 173, Minor, H. E. (ed.), VAW ETH Zurich, Zürich (in German).

Bollrich, G. (1996) *Technische Hydromechanik 1 - Grundlagen*. Verlag für Bauwesen, Berlin; München.

Bonet, J. and Kulasegaram, S. (2002) 'A simplified approach to enhance the performance of smooth particle hydrodynamics methods', *Applied Mathematics and Computation*, 126 (2-3), pp.133-155.

Borah, D. K., Alonso, C. V. and Prasad, S. N. (1982) 'Routing Graded Sediments in Streams - Formulations', *Journal of the Hydraulics Division-Asce*, 108 (12), pp.1486-1503.

Brendel, L. and Dippel, S. (1998) 'Lasting contacts in molecular dynamics simulations', *Physics of Dry Granular Media*, 350, pp.313-318.

Brilliantov, N. V. and Pöschel, T. (2004) *Kinetic theory of granular gases*. Oxford University Press, Oxford.

Brownlie, W. R. (1985) 'Compilation of alluvial channel data', *Journal of Hydraulic Engineering-Asce*, 111 (7), pp.1115-1119.

Brownlie, W. R. and Brooks, N. H. (1981) *Prediction of flow depth and sediment discharge in open channels*. W.M. Keck Laboratory of Hydraulics and Water Resources, California Institute of Technology.

Buffington, J. M. and Montgomery, D. R. (1997) 'A systematic analysis of eight decades of incipient motion studies, with special reference to gravel-bedded rivers', *Water Resources Research*, 33 (8), pp.1993-2029.

Bui, H. H., Sako, K. and Fukagawa, R. (2007) 'Numerical simulation of soil-water interaction using smoothed particle hydrodynamics (SPH) method', *Journal of Terramechanics*, 44 (5), pp.339-346.

Bui, M. D. and Rodi, W. (2008) 'Numerical simulation of contraction scour in an open laboratory channel', *Journal of Hydraulic Engineering-Asce*, 134 (4), pp.367-377.

Bui, M. D. and Rutschmann, P. (2010) 'Numerical modelling of non-equilibrium graded sediment transport in a curved open channel', *Computers & Geosciences*, 36 (6), pp.792-800.

Bungartz, H.-J., Mehl, M. and Schäfer, M. (eds.) (2010) *Fluid Structure Interaction II: Modelling, Simulation, Optimization*. Berlin; Heidelberg: Springer.

Burshtein, L. (1968) 'Determination of poisson's ratio for rocks by static and dynamic methods', *Journal of Mining Science*, 4 (3), pp.235-238.

Camenen, B. and Larson, M. (2005) 'A general formula for non-cohesive bed load sediment transport', *Estuarine, Coastal and Shelf Science*, 63 (1-2), pp.249-260.

Bibliography

Chaniotis, A. K., Poulikakos, D. and Koumoutsakos, P. (2002) 'Remeshed smoothed particle hydrodynamics for the simulation of viscous and heat conducting flows', *Journal of Computational Physics*, 182 (1), pp.67-90.

Chaudhry, M. (2008) *Open-Channel Flow*. Springer, New York.

Cheng, N.-S. (2004) 'Analysis of bedload transport in laminar flows', *Advances in Water Resources*, 27 (9), pp.937-942.

Cheng, N. S. and Chiew, Y. M. (1998) 'Pickup probability for sediment entrainment', *Journal of Hydraulic Engineering-Asce*, 124 (2), pp.232-235.

Chorin, A. J. (1967) 'Numerical solution of Navier-Stokes equations for an incompressible fluid', *Bulletin of the American Mathematical Society*, 73 (6), pp.928-931.

Chorin, A. J. (1968) 'Numerical solution of Navier-Stokes equations', *Mathematics of Computation*, 22 (104), pp.745-762.

Chorin, A. J. and Bernard, P. S. (1973) 'Discretization of a vortex sheet, with an example of roll-up', *Journal of Computational Physics*, 13 (3), pp.423-429.

Church, M. (2006) 'Bed material transport and the morphology of alluvial river channels', *Annual Review of Earth and Planetary Sciences*, 34, pp.325-354.

Cleary, P. W. and Prakash, M. (2004) 'Discrete-element modelling and smoothed particle hydrodynamics: potential in the environmental sciences', *Philosophical Transactions of the Royal Society a-Mathematical Physical and Engineering Sciences*, 362 (1822), pp.2003-2030.

Colagrossi, A., Colicchio, G., Lugni, C. and Brocchini, M. (2010) 'A study of violent sloshing wave impacts using an improved SPH method', *Journal of Hydraulic Research*, 48, pp.94-104.

Colagrossi, A. and Landrini, M. (2003) 'Numerical simulation of interfacial flows by smoothed particle hydrodynamics', *Journal of Computational Physics*, 191 (2), pp.448-475.

Coleman, N. L. (1972) 'The drag coefficient of a stationary sphere on boundary of similar spheres', *La Houille Blanche*, (1), pp.17-22.

Cook, B. K., Noble, D. R. and Williams, J. R. (2004) 'A direct simulation method for particle-fluid systems', *Engineering Computations*, 21 (2-4), pp.151-168.

Coquerelle, M. and Cottet, G. H. (2008) 'A vortex level set method for the two-way coupling of an incompressible fluid with colliding rigid bodies', *Journal of Computational Physics*, 227 (21), pp.9121-9137.

Cottet, G. H. (1996) 'Artificial viscosity models for vortex and particle methods', *Journal of Computational Physics*, 127 (2), pp.299-308.

Cottet, G. H. and Koumoutsakos, P. (2000) *Vortex Methods, Theroy and Practice*. Cambridge University Press, Cambridge.

Courant, R., Friedrichs, K. and Lewy, H. (1928) 'Partial differential equations of mathematical physics', *Mathematische Annalen*, 100, pp.32-74.

Crespo, A. J. C., Gomez-Gesteira, M. and Dalrymple, R. A. (2007) '3D SPH Simulation of large waves mitigation with a dike', *Journal of Hydraulic Research*, 45 (5), pp.631-642.

Cummins, S. J. and Brackbill, J. U. (2002) 'An implicit particle-in-cell method for granular materials', *Journal of Computational Physics*, 180 (2), pp.506-548.

Cummins, S. J. and Rudman, M. (1999) 'An SPH Projection Method', *Journal of Computational Physics*, 152 (2), pp.584-607.

Cundall, P. A. (1988) 'Formulation of a 3-dimensional distinct element model. 1. A scheme to detect and represent contacts in a system composed of many polyhedral blocks', *International Journal of Rock Mechanics and Mining Sciences & Geomechanics Abstracts*, 25 (3), pp.107-116.

Cundall, P. A. and Hart, R. D. (1992) 'Numerical Modelling of Discontinua', *Engineering Computations*, 9 (2), pp.101-113.

Cundall, P. A. and Strack, O. D. L. (1979) 'Discrete numerical-model for granular assemblies', *Geotechnique*, 29 (1), pp.47-65.

Cunge, J. A., Holly, F. M. J. and Verwey, A. (1980) *Practical aspects of computational river hydraulics.* Pitman, London.

D'Agostino, V. and Ferro, V. (2004) 'Scour on alluvial bed downstream of grade-control structures', *Journal of Hydraulic Engineering-Asce*, 130 (1), pp.24-37.

Dalrymple, R. A. and Rogers, B. D. (2006) 'Numerical modeling of water waves with the SPH method', *Coastal Engineering*, 53 (2-3), pp.141-147.

Dancey, C. L., Diplas, P., Papanicolaou, A. and Bala, M. (2002) 'Probability of individual grain movement and threshold condition', *Journal of Hydraulic Engineering-Asce*, 128 (12), pp.1069-1075.

Detert, M., Weitbrecht, V. and Jirka, G. H. (2010) 'Laboratory Measurements on Turbulent Pressure Fluctuations in and above Gravel Beds', *Journal of Hydraulic Engineering-Asce*, 136 (10), pp.779-789.

Dey, S. and Papanicolaou, A. (2008) 'Sediment Threshold under Stream Flow: A State-of-the-Art Review', *Ksce Journal of Civil Engineering*, 12 (1), pp.45-60.

DiFelice, R. (1996) 'A relationship for the wall effect on the settling velocity of a sphere at any flow regime', *Int. J. Multiph. Flow*, 22 (3), pp.527-533.

Do-Quang, M. and Amberg, G. (2010) 'Numerical simulation of the coupling problems of a solid sphere impacting on a liquid free surface', *Math. Comput. Simul.*, 80 (8), pp.1664-1673.

Douady, S., Andreotti, B. and Daerr, A. (1999) 'On granular surface flow equations', *European Physical Journal B*, 11 (1), pp.131-142.

Douglas, J. F., Gasiorek, J. M. and Swaffield, J. (2001) *Fluid mechanics.* 4th edn. Longmann, Harlow.

du Boys, P. (1879) 'Étude du régime du Rhone et de l'action exercée par les eaux sur un lit à fond de graviers indéfiniment affouillable (Study of flow regime of the Rhone and water force exerted on a gravel bed of infinite depth)', *Annales des Ponts et Chaussées*, 5 (19), pp.141-195.

Duez, C., Ybert, C., Clanet, C. and Bocquet, L. (2007) 'Making a splash with water repellency', *Nat Phys*, 3 (3), pp.180-183.

Dwivedi, A., Melville, B. and Shamseldin, A. Y. (2010) 'Hydrodynamic Forces Generated on a Spherical Sediment Particle during Entrainment', *Journal of Hydraulic Engineering-Asce*, 136 (10), pp.756-769.

Egiazaroff, I. W. (1965) 'Calculation of Non-Uniform Sediment Concentration', *Journal of Hydraulics Division, ASCE*, (44).

Einstein, H. A. (1937) 'Der Geschiebetrieb als Wahrscheinlichkeitsproblem', in Meyer-Peter, E. (ed.) *Mitteilung der Versuchsanstalt für Wasserbau an der Eidg. Techn. Hochschule in Zürich.* Rascher, Zürich, pp.3-112.

Einstein, H. A. (1950) 'The bed-load function for sediment transportation in open channel flow', *U S Dept Agric Tech Bull*, 1026, pp.1-71.

Einstein, H. A. and Elsamni, E. A. (1949) 'Hydrodynamic forces on a rough wall', *Reviews of Modern Physics*, 21 (3), pp.520-524.

Elhakeem, M. and Mran, J. (2007) 'Density functions for entrainment and deposition rates of uniform sediment', *Journal of Hydraulic Engineering-Asce*, 133 (8), pp.917-926.

Ellero, M., Serrano, M. and Espanol, P. (2007) 'Incompressible smoothed particle hydrodynamics', *Journal of Computational Physics*, 226, pp.1731-1752.

Ettema, R. and Mutel, C. F. (2004) 'Hans Albert Einstein: Innovation and Compromise in Formulating Sediment Transport by Rivers', *Journal of Hydraulic Engineering*, 130 (6), pp.477-487.

Exner, F. M. (1925) Ueber die Wechselwirkung zwischen Wasser und Geschiebe in Fluessen. Akademie der Wissenschaften, Mathematische Naturwissenschaften Abt. IIa, Wien, Austria.

Eymard, R., Gallouët, T. and Herbin, R. (2000) 'Finite volume methods', in Ciarlet, P. G. and Lions, J. L. (eds.) *Handbook of Numerical Analysis*. Elsevier, pp.713-1018.

Faeh, R., Mueller, R., Rousselot, P., Vetsch, D., Volz, C., Vonwiller, L., Veprek, R. and Farshi, D. (2011) System Manuals of BASEMENT, Version 2.1. Laboratory of Hydraulics, Glaciology and Hydrology (VAW), ETH Zurich, Zürich.

Falappi, S. and Gallati, M. (2007) 'SPH simulation of water waves generated by granular landslides', *32rd IAHR Congress*. Venice: p.107.

Fang, J., Parriaux, A., Rentschler, M. and Ancey, C. (2009) 'Improved SPH methods for simulating free surface flows of viscous fluids', *Applied Numerical Mathematics*, 59 (2), pp.251-271.

Feng, Y. T., Han, K. and Owen, D. R. J. (2007) 'Coupled lattice Boltzmann method and discrete element modelling of particle transport in turbulent fluid flows: Computational issues', *International Journal for Numerical Methods in Engineering*, 72, pp.1111-1134.

Feng, Y. T., Han, K. and Owen, D. R. J. (2010) 'Combined three-dimensional lattice Boltzmann method and discrete element method for modelling fluid-particle interactions with experimental assessment', *International Journal for Numerical Methods in Engineering*, 81 (2), pp.229-245.

Fenton, J. D. and Abbott, J. E. (1977) 'Initial movement of grains on a stream bed - effect of relative protrution', *Proceedings of the Royal Society of London Series a-Mathematical Physical and Engineering Sciences*, 352 (1671), pp.523-537.

Fincham, D. (1992) 'Leapfrog rotational algorithms', *Molecular Simulation*, 8 (3-5), pp.165-178.

Fleissner, F. (2010) *Parallel Object Oriented Simulation with Lagrangian Particle Methods*. Shaker, Aachen.

Fleissner, F., D'Alessandro, V., Schiehlen, W. and Eberhard, P. (2009a) 'Sloshing cargo in silo vehicles', *Journal of Mechanical Science and Technology*, 23 (4), pp.968-973.

Fleissner, F., Gaugele, T. and Eberhard, P. (2007) 'Applications of the discrete element method in mechanical engineering', *Multibody System Dynamics*, 18 (1), pp.81-94.

Fleissner, F., Haag, T., Hanss, M. and Eberhard, P. (2009b) 'Uncertainty Analysis for a Particle Model of Granular Chute Flow', *Cmes-Computer Modeling in Engineering & Sciences*, 52 (2), pp.181-196.

Fleissner, F., Lehnart, A. and Eberhard, P. (2010) 'Dynamic simulation of sloshing fluid and granular cargo in transport vehicles', *Vehicle System Dynamics*, 48 (1), pp.3-15.

Francalanci, S., Parker, G. and Solari, L. (2008) 'Effect of seepage-induced nonhydrostatic pressure distribution on bed-load transport and bed morphodynamics', *Journal of Hydraulic Engineering-Asce*, 134 (4), pp.378-389.

Gaugele, T., Fleissner, F. and Eberhard, P. (2008) 'Simulation of material tests using meshfree Lagrangian particle methods', *Proceedings of the Institution of Mechanical Engineers Part K-Journal of Multi-Body Dynamics*, 222 (4), pp.327-338.

Gessler, D., Hall, B., Spasojevic, M., Holly, F., Pourtaheri, H. and Rapheli, N. (1999) 'Application of 3D mobile bed, hydrodynamic model', *Journal of Hydraulic Engineering-Asce*, 125 (7), pp.737-749.

Gessler, J. (1965) Der Geschiebetriebbeginn bei Mischungen untersucht an natürlichen Abpfästerungserscheinungen in Kanälen. *Mitteilungen der Versuchsanstalt für Wasserbau und Erdbau* 69, Schnitter, G. (ed.), VAW ETH Zurich, Zürich (in German).

Gingold, R. A. and Monaghan, J. J. (1977) 'Smoothed Particle Hydrodynamics - theory and applications to non-spherical stars', *Mon. Not. Roy. Astron. Soc.*, 181 (2), pp.375-389.

Bibliography

Glowinski, R. (2003) 'Finite element methods for incompressible viscous flow', in Ciarlet, P. G. and Lions, J. L. (eds.) *Handbook of Numerical Analysis*. Elsevier, pp.3-1176.

Gomez-Gesteira, M., Rogers, B. D., Dalrymple, R. A. and Crespo, A. J. C. (2010a) 'State-of-the-art of classical SPH for free-surface flows', *Journal of Hydraulic Research*, 48 (1 supp 1), pp.6 - 27.

Gomez-Gesteira, M., Rogers, B. D., Violeau, D., Grassa, J. M. and Crespo, A. J. C. (2010b) 'Foreword: SPH for free-surface flows', *Journal of Hydraulic Research*, 48, pp.3-5.

Graf, W. H. (1971) *Hydraulics of Sediment Transport*. McGraw-Hill, New York.

Graham, D. I. and Hughes, J. P. (2008) 'Accuracy of SPH viscous flow models', *International Journal for Numerical Methods in Fluids*, 56 (8), pp.1261-1269.

Griebel, M., Knapek, S. and Zumbusch, G. (2007) *Numerical simulation in molecular dynamics*. Springer, Berlin.

Günter, A. (1971) Die kritische mittlere Sohlenschubspannung bei Geschiebemischungen unter Berücksichtigung der Deckschichtbildung und der turbulenzbedingten Sohlenschubspannungsschwankungen. *VAW Mitteilung* 3, Vischer, D. (ed.), VAW ETH Zurich, Zürich (in German).

Habersack, H. M. and Laronne, J. B. (2002) 'Evaluation and improvement of bed load discharge formulas based on Helley-Smith sampling in an alpine gravel bed river', *Journal of Hydraulic Engineering-Asce*, 128 (5), pp.484-499.

Harlow, F. H. (1964) *The particle-in-cell computing method for fluid dynamics*. Academic Press, New York.

Harlow, F. H. and Welch, J. E. (1965) 'Numerical Calculation of time-dependent viscous incompressible flow of fluid with free surface', *Physics of Fluids*, 8 (12), pp.2182-2190.

Hart, R., Cundall, P. A. and Lemos, J. (1988) 'Formulation of a 3-dimensional distinct element model. 2. Mechanical calculations for motion and interaction of a system composed of many polyhedral blocks', *International Journal of Rock Mechanics and Mining Sciences & Geomechanics Abstracts*, 25 (3), pp.117-125.

Hernquist, L. and Katz, N. (1989) 'TREESPH - A unification of SPH with the hierarchical tree method', *Astrophysical Journal Supplement Series*, 70 (2), pp.419-446.

Herrmann, H. J., Andrade, J. S., Araujo, A. D. and Almeida, M. P. (2007) 'Particles in fluids', *European Physical Journal-Special Topics*, 143, pp.181-189.

Herrmann, H. J. and Luding, S. (1998) 'Modeling granular media on the computer', *Continuum Mechanics and Thermodynamics*, 10 (4), pp.189-231.

Hertz, H. (1882) 'Ueber die Berührung fester elastischer Körper', *Journal für die reine und angewandte Mathematik (Crelles Journal)*, 1882 (92), pp.156-171.

Hieber, S. E. and Koumoutsakos, P. (2008) 'An immersed boundary method for smoothed particle hydrodynamics of self-propelled swimmers', *Journal of Computational Physics*, 227 (19), pp.8636-8654.

Hirsch, C. (1988) *Numerical computation of internal and external flows*. Jon Wiley & Sons, New York.

Hirt, C. W. (1993) 'Volume-fraction techniques - powerful tools for wind engineering', *Journal of Wind Engineering and Industrial Aerodynamics*, 46-7, pp.327-338.

Hirt, C. W. and Nichols, B. D. (1981) 'Volume of Fluid (VOF) Method for the dynamics of Free Boundaries', *Journal of Computational Physics*, 39 (1), pp.201-225.

Hofland, B. and Battjes, J. A. (2006) 'Probability density function of instantaneous drag forces and shear stresses on a bed', *Journal of Hydraulic Engineering-Asce*, 132 (11), pp.1169-1175.

Holger, W. (2009) *Scattered data approximation*. Cambridge University Press, Cambridge.

Huerta, A., Belytschko, T., Fernández-Méndez, S. and Rabczuk, T. (2004) 'Meshfree Methods', in *Encyclopedia of Computational Mechanics*. John Wiley & Sons, New York.

Hughes, J. P. and Graham, D. I. (2010) 'Comparison of incompressible and weakly-compressible SPH models for free-surface water flows', *Journal of Hydraulic Research*, 48, pp.105-117.

Hunziker, R. P. (1995) Fraktionsweiser Geschiebetransport. *VAW Mitteilung* 138, Vischer, D. (ed.), VAW ETH Zurich, Zürich.

Hunziker, R. P. and Jaeggi, M. N. R. (2002) 'Grain sorting processes', *Journal of Hydraulic Engineering-Asce*, 128 (12), pp.1060-1068.

Hutter, K. and Jöhnk, K. (2004) *Continuum methods of physical modeling: continuum mechanics, dimensional analysis, turbulence.* Springer, Berlin.

Hutter, K. and Rajagopal, K. R. (1994) 'On flows of granular-materials', *Continuum Mechanics and Thermodynamics*, 6 (2), pp.81-139.

Hutter, K. and Schneider, L. (2010a) 'Important aspects in the formulation of solid-fluid debris-flow models. Part I. Thermodynamic implications', *Continuum Mechanics and Thermodynamics*, 22 (5), pp.363-390.

Hutter, K. and Schneider, L. (2010b) 'Important aspects in the formulation of solid-fluid debris-flow models. Part II. Constitutive modelling', *Continuum Mechanics and Thermodynamics*, 22 (5), pp.391-411.

Ikeda, S. (1982) 'Lateral Bed-Load Transport on Side Slopes', *Journal of the Hydraulics Division-Asce*, 108 (11), pp.1369-1373.

Issa, N. (2005) Numerical assessment of the smoothed particle hydrodynamics gridless method for incompressible flows and its extension to turbulent flows. *PhD Thesis*. Institute of Science and Technology, University of Manchester, Manchester.

Issa, R., Lee, E. S., Violeau, D. and Laurence, D. R. (2005) 'Incompressible separated flows simulations with the smoothed particle hydrodynamics gridless method', *International Journal for Numerical Methods in Fluids*, 47 (10-11), pp.1101-1106.

Iverson, R. M. (1997) 'The physics of debris flows', *Reviews of Geophysics*, 35 (3), pp.245-296.

Jain, S. C. (1992) 'Note on lag in bedload discharge', *Journal of Hydraulic Engineering-Asce*, 118 (6), pp.904-917.

Jang, Y., Biddiscombe, J., Marongiu, J.-C. and Parkinson, E. (2009) Unified Visual Post Processing of Mesh and Meshless Data for Hydrodynamic Simulations. *Scientific Report*, Swiss National Supercomputing Center (CSCS), Manno.

Jang, Y., Fuchs, R., Schindler, B. and Peikert, R. (2010) 'Volumetric Evaluation of Meshless Data From Smoothed Particle Hydrodynamics Simulations', *IEEE/EG International Symposium on Volume Graphics*.

Jayaweer, K. O. L. F. and Mason, B. J. (1965) 'Behaviour of freely falling cylinders and cones in a viscous fluid', *Journal of Fluid Mechanics*, 22, pp.709-718.

Jiang, Z. and Haff, P. K. (1993) 'Multiparticle simulation methods applied to the micromechanics of bed-load transport', *Water Resources Research*, 29 (2), pp.399-412.

Kacianauskas, R., Balevicius, R., Markauskas, D. and Maknickas, A. (2007) 'Discrete element method in simulation of granular materials', *IUTAM Symposium on Multiscale Problems in Multibody System Contacts*, 1, pp.65-74.

Karamanev, D. G. (1996) 'Equations for calculation of the terminal velocity and drag coefficient of solid spheres and gas bubbles', *Chemical Engineering Communications*, 147, pp.75-84.

Bibliography

Kawaguchi, T., Tanaka, T. and Tsuji, Y. (1998) 'Numerical simulation of two-dimensional fluidized beds using the discrete element method (comparison between the two- and three-dimensional models)', *Powder Technology*, 96 (2), pp.129-138.

Kern, S. and Koumoutsakos, P. (2006) 'Simulations of optimized anguilliform swimming', *Journal of Experimental Biology*, 209 (24), pp.4841-4857.

Kleinhans, M. G. and van Rijn, L. C. (2002) 'Stochastic prediction of sediment transport in sand-gravel bed rivers', *Journal of Hydraulic Engineering-Asce*, 128 (4), pp.412-425.

Knudsen, J. M. and Hjorth, P. G. (2000) *Elements of Newtonian mechanics.* Springer, Berlin.

Koumoutsakos, P. (2005) 'Multiscale flow simulations using particles', *Annual Review of Fluid Mechanics*, 37, pp.457-487.

Koumoutsakos, P., Leonard, A. and Pepin, F. (1994) 'Boundary-conditions for viscous vortex methods', *Journal of Computational Physics*, 113 (1), pp.52-61.

Krištof, P., Beneš, B., Křivánek, J. and Šťava, O. (2009) 'Hydraulic Erosion Using Smoothed Particle Hydrodynamics', *Computer Graphics Forum*, 28 (2), pp.219-228.

Landrini, M., Colagrossi, A., Greco, M. and Tulin, M. P. (2007) 'Gridless simulations of splashing processes and near-shore bore propagation', *Journal of Fluid Mechanics*, 591, pp.183-213.

Lankarani, H. M. and Nikravesh, P. (1994) 'Continuous contact force models for impact analysis in multibody systems', *Nonlinear Dynamics*, 5 (2), pp.193-207.

Lanru, J. and Ove, S. (2007) *Fundamentals of discrete element methods for rock engineering : theory and applications.* Elsevier, Amsterdam

Lee, E.-S., Violeau, D., Issa, R. and Ploix, S. (2010) 'Application of weakly compressible and truly incompressible SPH to 3-D water collapse in waterworks', *Journal of Hydraulic Research*, 48 (1 supp 1), pp.50 - 60.

Lee, E. S., Moulinec, C., Xu, R., Violeau, D., Laurence, D. and Stansby, P. (2008) 'Comparisons of weakly compressible and truly incompressible algorithms for the SPH mesh free particle method', *Journal of Computational Physics*, 227 (18), pp.8417-8436.

Lehnart, A. (2008) Ein Smoothed Particle Hydrodynamics-Verfahren zur Behandlung der Eulergleichungen. *MSc Thesis.* Institut für Technische und Numerische Mechanik, Universität Stuttgart, Stuttgart.

Leonard, A. (1980) 'Vortex methods for flow simulation', *Journal of Computational Physics*, 37 (3), pp.289-335.

Lesser, G. R., Roelvink, J. A., van Kester, J. A. T. M. and Stelling, G. S. (2004) 'Development and validation of a three-dimensional morphological model', *Coastal Engineering*, 51 (8-9), pp.883-915.

Libersky, L. D., Petschek, A. G., Carney, T. C., Hipp, J. R. and Allahdadi, F. A. (1993) 'High-strain Lagrangian hydrodynamics - A 3-dimensional SPH code for dynamic material response', *Journal of Computational Physics*, 109 (1), pp.67-75.

Liu, G. R. and Liu, M. B. (2003) *Smoothed particle hydrodynamics: a meshfree particle method.* World Scientific, New Jersey.

Liu, M. B. and Liu, G. R. (2010) 'Smoothed Particle Hydrodynamics (SPH): an Overview and Recent Developments', *Archives of Computational Methods in Engineering*, 17 (1), pp.25-76.

Lucy, L. B. (1977) 'Numerical approach to testing of fission hypothesis', *Astron. J.*, 82 (12), pp.1013-1024.

Luding, S. (1998) 'Collisions & contacts between two particles', *Physics of Dry Granular Media*, 350, pp.285-304.

Manenti, S., Agate, G., Di Monaco, A., Gallati, M., Maffio, A., Guandalini, R. and Sibilla, S. (2009) 'SPH Modeling of Rapid Sediment Scour Induced by Water Flow', *33rd IAHR Congress.* Vancouver.

Mantz, P. A. (1977) 'Incipient Transport of Fine Grains and Flakes by Fluids — Extended Shields Diagram', *Journal of the Hydraulics Division-Asce*, 103 (6), pp.601-615.

Maruzewski, P., Touzé, D. L., Oger, G. and Avellan, F. (2010) 'SPH high-performance computing simulations of rigid solids impacting the free-surface of water', *Journal of Hydraulic Research*, 48 (1 supp 1), pp.126-134.

McEwan, I. and Heald, J. (2001) 'Discrete particle modeling of entrainment from flat uniformly sized sediment beds', *Journal of Hydraulic Engineering-Asce*, 127 (7), pp.588-597.

McKee, S., Tome, M. F., Cuminato, J. A., Castelo, A. and Ferreira, V. G. (2004) 'Recent advances in the marker and cell method', *Archives of Computational Methods in Engineering*, 11 (2), pp.107-142.

Meyer-Peter, E., Favre, H. and Einstein, H. A. (1934) 'Neuere Versuchsresultate über den Geschiebetrieb', *Schweizer Bauzeitung*, 103 (13).

Meyer-Peter, E. and Müller, R. (1948) 'Formulas for Bed-Load Transport', *2nd Meeting IAHR*. Stockholm, Sweden: IAHR, pp.39-64.

Mitchell, J. K. and Soga, K. (2005) *Fundamentals of Soil Behavior*. 3rd edn. John Wiley & Sons.

Mittal, R. and Iaccarino, G. (2005) 'Immersed boundary methods', *Annual Review of Fluid Mechanics*, 37, pp.239-261.

Mohamad, A. (2011) *Lattice Boltzmann method : fundamentals and engineering applications with computer codes*. London : Springer.

Molteni, D. and Colagrossi, A. (2009) 'A simple procedure to improve the pressure evaluation in hydrodynamic context using the SPH', *Computer Physics Communications*, 180 (6), pp.861-872.

Monaghan, J. J. (1985) 'Particle methods for hydrodynamics', *Computer Physics reports*, 3 (2), pp.71-124.

Monaghan, J. J. (1989) 'On the problem of penetration in particle methods', *Journal of Computational Physics*, 82 (1), pp.1-15.

Monaghan, J. J. (1992) 'Smoothed Particle Hydrodynamics', *Annu. Rev. Astron. Astrophys.*, 30, pp.543-574.

Monaghan, J. J. (1994) 'Simulating free-surface flows with SPH', *Journal of Computational Physics*, 110 (2), pp.399-406.

Monaghan, J. J. (2000) 'SPH without a tensile instability', *Journal of Computational Physics*, 159 (2), pp.290-311.

Monaghan, J. J. (2002) 'SPH compressible turbulence', *Mon. Not. Roy. Astron. Soc.*, 335 (3), pp.843-852.

Monaghan, J. J. (2005a) 'Smoothed particle hydrodynamics', *Reports on Progress in Physics*, 68 (8), pp.1703-1759.

Monaghan, J. J. (2005b) 'Theory and Applications of Smoothed Particle Hydrodynamics', in Blowey, J. F. and Craig, A. W. (eds.) *Frontiers of Numerical Analysis*. Springer Berlin Heidelberg, pp.143-194.

Monaghan, J. J. (2006) 'Smoothed particle hydrodynamic simulations of shear flow', *Mon. Not. Roy. Astron. Soc.*, 365 (1), pp.199-213.

Monaghan, J. J., Cas, R. A. F., Kos, A. M. and Hallworth, M. (1999) 'Gravity currents descending a ramp in a stratified tank', *Journal of Fluid Mechanics*, 379, pp.39-70.

Monaghan, J. J. and Gingold, R. A. (1983) 'Shock simulation by the particle method SPH', *Journal of Computational Physics*, 52 (2), pp.374-389.

Monaghan, J. J. and Kajtar, J. B. (2009) 'SPH particle boundary forces for arbitrary boundaries', *Computer Physics Communications*, 180 (10), pp.1811-1820.

Monaghan, J. J. and Kocharyan, A. (1995) 'SPH simulation of multiphase flow', *Computer Physics Communications*, 87 (1-2), pp.225-235.

Bibliography

Monaghan, J. J. and Kos, A. (1999) 'Solitary waves on a Cretan beach', *Journal of Waterway Port Coastal and Ocean Engineering-Asce*, 125 (3), pp.145-154.

Monaghan, J. J. and Kos, A. (2000) 'Scott Russell's wave generator', *Physics of Fluids*, 12 (3), pp.622-630.

Monaghan, J. J., Kos, A. and Issa, N. (2003) 'Fluid motion generated by impact', *Journal of Waterway Port Coastal and Ocean Engineering-Asce*, 129 (6), pp.250-259.

Monaghan, J. J. and Lattanzio, J. C. (1986) 'Further-studies of a fragmentation problem', *Astronomy and Astrophysics*, 158 (1-2), pp.207-211.

Morris, J. P., Fox, P. J. and Zhu, Y. (1997) 'Modeling low Reynolds number incompressible flows using SPH', *Journal of Computational Physics*, 136 (1), pp.214-226.

Muller, M., Charypar, D. and Gross, M. (2003) 'Particle-based fluid simulation for interactive applications', *ACM SIGGRAPH/Eurographics Symposium on Computer Animation*. 26-27 July. San Diego, CA,: Assoc. for Comput. Machinery, pp.154-159.

Muller, M., Schirm, S., Teschner, M., Heidelberger, B. and Gross, M. (2004) 'Interaction of fluids with deformable solids', *Computer Animation and Virtual Worlds*, 15 (3-4), pp.159-171.

Münsch, M. and Breuer, M. (2010) 'Numerical Simulation of Fluid–Structure Interaction Using Eddy–Resolving Schemes', in Bungartz, H.-J., Mehl, M. and Schäfer, M. (eds.) *Fluid Structure Interaction II*. Springer, Berlin, pp.221-253.

Murphy, P. J. and Aguirre, E. J. (1985) 'Bed-load or suspended-load', *Journal of Hydraulic Engineering-Asce*, 111 (1), pp.93-107.

Murray, A. B. and Paola, C. (1994) 'A cellular-model of braided rivers', *Nature*, 371 (6492), pp.54-57.

Nguyen, V. P., Rabczuk, T., Bordas, S. and Duflot, M. (2008) 'Meshless methods: A review and computer implementation aspects', *Math. Comput. Simul.*, 79 (3), pp.763-813.

Nikuradse, J. (1933) 'Flow laws in raised tubes', *Zeitschrift Des Vereines Deutscher Ingenieure*, 77, pp.1075-1076.

Olsen, N. R. B. (2003) 'Three-dimensional CFD modeling of self-forming meandering channel', *Journal of Hydraulic Engineering-Asce*, 129 (5), pp.366-372.

Omelyan, I. P. (1998) 'On the numerical integration of motion for rigid polyatomics: The modified quaternion approach', *Computers in Physics*, 12 (1), pp.97-103.

Onate, E., Idelsohn, S. R., Celigueta, M. A. and Rossi, R. (2008) 'Advances in the particle finite element method for the analysis of fluid-multibody interaction and bed erosion in free surface flows', *Computer Methods in Applied Mechanics and Engineering*, 197 (19-20), pp.1777-1800.

Onate, E., Idelsohn, S. R., Del Pin, F. and Aubry, R. (2004) 'The particle finite element method - an overview', *International Journal of Computational Methods*, 1 (2), pp.267-307.

Owen, D. R. J., Leonardi, C. R. and Feng, Y. T. (2011) 'An efficient framework for fluid–structure interaction using the lattice Boltzmann method and immersed moving boundaries', *International Journal for Numerical Methods in Engineering*, 87 (1-5), pp.66-95.

Paintal, A. S. (1971) 'Concept of Critical Shear Stress in Loose Boundary Open Channels', *Journal of Hydraulic Research*, 9 (1), pp.91-114.

Panton, R. L. (2005) *Incompressible flow*. Jon Wiley & Sons, New York.

Paola, C. and Voller, V. R. (2005) 'A generalized Exner equation for sediment mass balance', *J. Geophys. Res.*, 110 (F4), p.F04014.

Papanicolaou, A. N. (1999) 'Pickup probability for sediment entrainment - Discussion', *Journal of Hydraulic Engineering-Asce*, 125 (7), pp.788-789.

Papanicolaou, A. N., Diplas, P., Evaggelopoulos, N. and Fotopoulos, S. (2002) 'Stochastic incipient motion criterion for spheres under various bed packing conditions', *Journal of Hydraulic Engineering-Asce*, 128 (4), pp.369-380.

Paphitis, D. (2001) 'Sediment movement under unidirectional flows: an assessment of empirical threshold curves', *Coastal Engineering*, 43 (3-4), pp.227-245.

Parker, G. (1990) 'Surface-based bedload transport relation for gravel rivers', *Journal of Hydraulic Research*, 28 (4), pp.417-436.

Parker, G., Paola, C. and Leclair, S. (2000) 'Probabilistic Exner sediment continuity equation for mixtures with no active layer', *Journal of Hydraulic Engineering-Asce*, 126 (11), pp.818-826.

Peskin, C. S. (1972) 'Flow patterns around heart valves - numerical method', *Journal of Computational Physics*, 10 (2), pp.252-271.

Peskin, C. S. (1977) 'Numerical analysis of blood flow in the heart', *Journal of Computational Physics*, 25 (3), pp.220-252.

Phillips, B. C. and Sutherland, A. J. (1989) 'Spatial lag effects in bed-load sediment transport', *Journal of Hydraulic Research*, 27 (1), pp.115-133.

Pilotti, M. and Menduni, G. (2001) 'Beginning of sediment transport of incoherent grains in shallow shear flows', *Journal of Hydraulic Research*, 39 (2), pp.115-124.

Popov, V. L. (2010) *Contact Mechanics and Friction : Physical Principles and Applications.* Berlin : Springer.

Pöschel, T. and Schwager, T. (2005) *Computational granular dynamics.* Springer, Berlin.

Potapov, A. V. and Campbell, C. S. (1996) 'Computer simulation of hopper flow', *Physics of Fluids*, 8 (11), pp.2884-2894.

Potapov, A. V., Hunt, M. L. and Campbell, C. S. (2001) 'Liquid-solid flows using smoothed particle hydrodynamics and the discrete element method', *Powder Technology*, 116 (2-3), pp.204-213.

Powell, M. S. P. M. S., Weerasekara, N. S., Cole, S., LaRoche, R. D. and Favier, J. (2011) 'DEM modelling of liner evolution and its influence on grinding rate in ball mills', *Miner. Eng.*, 24 (3-4), pp.341-351.

Price, D. J. (2004) Magnetic fields in Astrophysics. Institute of Astronomy & Churchill College, Cambridge, Cambridge.

Pudasaini, S. P. and Hutter, K. (2007) *Avalanche dynamics : dynamics of rapid flows of dense granular avalanches.* Springer, Berlin.

Qiu, L. C. (2008) 'Two-dimensional SPH simulations of landslide-generated water waves', *Journal of Hydraulic Engineering-Asce*, 134 (5), pp.668-671.

Raad, P. E. and Bidoae, R. (2005) 'The three-dimensional Eulerian-Lagrangian marker and micro cell method for the simulation of free surface flows', *Journal of Computational Physics*, 203 (2), pp.668-699.

Radice, A., Malavasi, S. and Ballio, F. (2006) 'Solid transport measurements through image processing', *Experiments in Fluids*, 41 (5), pp.721-734.

Radice, A., Malavasi, S. and Ballio, F. (2008) 'Sediment kinematics in abutment scour', *Journal of Hydraulic Engineering-Asce*, 134 (2), pp.146-156.

Rannacher, R. and Richter, T. (2010) 'An Adaptive Finite Element Method for Fluid-Structure Interaction Problems Based on a Fully Eulerian Formulation', in Bungartz, H.-J., Mehl, M. and Schäfer, M. (eds.) *Fluid Structure Interaction II.* Springer, Berlin, pp.159-191.

Rasio, F. A. (2000) 'Particle Methods in Astrophysical Fluid Dynamics', *Progress of Theoretical Physics Supplement*, 138 (Copyright (c) Progress of Theoretical Physics 2000 All rights reserved.), p.609.

Recking, A. (2009) 'Theoretical development on the effects of changing flow hydraulics on incipient bed load motion', *Water Resources Research*, 45.

Rickenmann, D. (1991) 'Hyperconcentrated flow and sediment transport at steep slopes', *Journal of Hydraulic Engineering-Asce*, 117 (11), pp.1419-1439.

Rickenmann, D. (1999) 'Empirical relationships for debris flows', *Natural Hazards*, 19 (1), pp.47-77.

Rider, W. J. and Kothe, D. B. (1998) 'Reconstructing volume tracking', *Journal of Computational Physics*, 141 (2), pp.112-152.

Roache, P. J. (1998) *Computational fluid dynamics.* Hermosa Publishers, Albuquerque - N.M.

Rodi, W. (1995) 'Impact of Reynolds-average modeling in hydraulics', *Proceedings of the Royal Society of London Series a-Mathematical and Physical Sciences*, 451 (1941), pp.141-164.

Rodi, W. (2006) 'DNS and LES of some engineering flows', *Fluid Dynamics Research*, 38 (2-3), pp.145-173.

Rogers, B. D. and Dalrymple, R. A. (2008) 'SPH modeling of Tsunami waves', *Advanced Numerical Models for Simulating Tsunami Waves and Runup*, 10, pp.75-100.

Roshko, A. (1961) 'Experiments on the flow past a circular cylinder at very high Reynolds number', *Journal of Fluid Mechanics*, 10 (3), pp.345-356.

Santamarina, J. C. and Cho, G. C. (2004) 'Soil behaviour: The role of particle shape', *Advances in Geotechnical Engineering: The Skempton Conference.* pp.604-617.

Savage, S. B. (1984) 'The mechanics of rapid granular flows', *Advances in Applied Mechanics*, 24, pp.289-366.

Savage, S. B. and Hutter, K. (1989) 'The motion of a finte mass of granular material down a rough incline', *Journal of Fluid Mechanics*, 199, pp.177-215.

Schellart, W. P. (2000) 'Shear test results for cohesion and friction coefficients for different granular materials: scaling implications for their usage in analogue modelling', *Tectonophysics*, 324 (1-2), pp.1-16.

Schlichting, H. (1936) 'Experimentelle Untersuchungen zum Rauhigkeitsproblem', *Archive of Applied Mechanics*, 7 (1), pp.1-34.

Schlichting, H., Gersten, K. and Krause, E. (2000) *Boundary-layer theory.* 8th edn. Springer, Berlin.

Schmeeckle, M. W. and Nelson, J. M. (2003) 'Direct numerical simulation of bedload transport using a local, dynamic boundary condition', *Sedimentology*, 50 (2), pp.279-301.

Schmeeckle, M. W., Nelson, J. M. and Shreve, R. L. (2007) 'Forces on stationary particles in near-bed turbulent flows', *Journal of Geophysical Research-Earth Surface*, 112 (F2).

Schoklitsch, A. (1926) *Die Geschiebebewegung an Flüssen und Stauwerken.* Springer, Vienna.

Schwammle, V. and Herrmann, H. J. (2003) 'Geomorphology: Solitary wave behaviour of sand dunes - Colliding dunes appear to traverse through one another and emerge unscathed', *Nature*, 426 (6967), pp.619-620.

Sethian, J. A. and Smereka, P. (2003) 'Level set methods for fluid interfaces', *Annual Review of Fluid Mechanics*, 35, pp.341-372.

Shabana, A. A. (2010) *Computational dynamics.* 3rd edn. John Wiley & Sons, Chichester.

Shakibaeinia, A. and Jin, Y. C. (2011) 'A mesh-free particle model for simulation of mobile-bed dam break', *Advances in Water Resources*, 34 (6), pp.794-807.

Shao, S. and Gotoh, H. (2005) 'Turbulence particle models for tracking free surfaces', *Journal of Hydraulic Research*, 43 (3), pp.276-289.

Shao, S. D. (2006) 'Simulation of breaking wave by SPH method coupled with k-epsilon model', *Journal of Hydraulic Research*, 44 (3), pp.338-349.

Shao, S. D. (2009) 'Incompressible SPH simulation of water entry of a free-falling object', *International Journal for Numerical Methods in Fluids*, 59 (1), pp.91-115.

Shao, S. D. and Lo, E. Y. M. (2003) 'Incompressible SPH method for simulating Newtonian and non-Newtonian flows with a free surface', *Advances in Water Resources*, 26 (7), pp.787-800.

Shields, A. (1936) Anwendungen der Ähnlichkeitsmechanik und der Turbulenzforschung auf die Geschiebebewegungen. *Mitteilungen der Preussische Versuchsanstalt für Wasserbau und Schiffbau* Heft 26, Preussische Versuchsanstalt für Wasserbau und Schiffbau, Berlin (in German).

Shu, A. P. and Fei, X. J. (2008) 'Sediment transport capacity of hyperconcentrated flow', *Science in China Series G-Physics Mechanics & Astronomy*, 51 (8), pp.961-975.

Smart, G. M. and Habersack, H. M. (2007) 'Pressure fluctuations and gravel entrainment in rivers', *Journal of Hydraulic Research*, 45 (5), pp.661-673.

Smart, G. M. and Jäggi, M. N. R. (1983) Sedimenttransport in steilen Gerinnen. *VAW Mitteilung* 64, Vischer, D. (ed.), VAW ETH Zurich, Zürich (in German).

Snider, D. M., O'Rourke, P. J. and Andrews, M. J. (1998) 'Sediment flow in inclined vessels calculated using a multiphase particle-in-cell model for dense particle flows', *Int. J. Multiph. Flow*, 24 (8), pp.1359-1382.

Spalart, P. R. (2009) 'Detached-Eddy Simulation', *Annual Review of Fluid Mechanics*, 41, pp.181-202.

Stam, J. and Fiume, E. (1995) 'Depicting fire and other gaseous phenomena using diffusion processes', *Computer Graphics Proceedings. SIGGRAPH 95*.

Sukop, M. C. and Thorne, D. T. (2006) *Lattice Boltzmann modeling: an introduction for geoscientists and engineers*. Springer, Berlin.

Sukumaran, B. and Ashmawy, A. K. (2001) 'Quantitative characterisation of the geometry of discrete particles', *Geotechnique*, 51 (7), pp.619-627.

Sun, Z. L. and Donahue, J. (2000) 'Statistically derived bedload formula for any fraction of nonuniform sediment', *Journal of Hydraulic Engineering-Asce*, 126 (2), pp.105-111.

Tavarez, F. A. and Plesha, M. E. (2007) 'Discrete element method for modelling solid and particulate materials', *International Journal for Numerical Methods in Engineering*, 70 (4), pp.379-404.

Teufelsbauer, H., Wang, Y., Chiou, M. C. and Wu, W. (2009) 'Flow-obstacle interaction in rapid granular avalanches: DEM simulation and comparison with experiment', *Granular Matter*, 11 (4), pp.209-220.

Teufelsbauer, H., Wang, Y., Pudasaini, S. P., Borja, R. I. and Wu, W. (2011) 'DEM simulation of impact force exerted by granular flow on rigid structures', *Acta Geotechnica*, 6 (3), pp.119-133.

Thomas, R., Nicholas, A. P. and Quine, T. A. (2007) 'Cellular modelling as a tool for interpreting historic braided river evolution', *Geomorphology*, 90 (3-4), pp.302-317.

Tognacca, C., Bezzola, G. R. and Minor, H. E. (2000) 'Threshold criterion for debris-flow initiation due to channel-bed failure', *Debris-Flow Hazards Mitigation: Mechanics, Prediction, and Assessment*, pp.89-97.

Toro, E. F. (1997) *Riemann Solvers and Numerical Methods for Fluid Dynamics*. 2nd edn. Springer, Berlin.

Tsuji, Y., Kawaguchi, T. and Tanaka, T. (1993) 'Discrete particle simulation of two-dimensional fluidized bed', *Powder Technology*, 77 (1), pp.79-87.

Tsuji, Y., Tanaka, T. and Ishida, T. (1992) 'Lagrangian numerical-simulation of plug flow of cohesionless particles in a hoizontal pipe', *Powder Technology*, 71 (3), pp.239-250.

Bibliography

Turton, R. and Levenspiel, O. (1986) 'A short note on the drag correlation for spheres', *Powder Technology*, 47 (1), pp.83-86.

Unger, J. and Hager, W. (2007) 'Down-flow and horseshoe vortex characteristics of sediment embedded bridge piers', *Experiments in Fluids*, 42 (1), pp.1-19.

Valyrakis, M., Diplas, P., Dancey, C. L., Greer, K. and Celik, A. O. (2010) 'Role of instantaneous force magnitude and duration on particle entrainment', *Journal of Geophysical Research-Earth Surface*, 115.

van Rijn, L. C. (1984) 'Sediment Transport, Part II: Suspended Load Transport', *Journal of Hydraulic Engineering, ASCE*, 110 (11).

Van Rijn, L. C. (1987) Mathematical modelling of morphological processes in the case of suspended sediment transport. Delft hydraulics communication No. 382,

van Rijn, L. C. (2007) 'Unified view of sediment transport by currents and waves. I: Initiation of motion, bed roughness, and bed-load transport', *Journal of Hydraulic Engineering-Asce*, 133 (6), pp.649-667.

Verlet, L. (1967) 'Computer experiments on classical fluids. I. Thermodynamical properties of Lennard-Jones molecules', *Physical Review*, 159 (1), pp.98-103.

Vince, J. (2008) *Geometric algebra for computer graphics*. Springer, London.

Violeau, D. and Issa, R. (2007a) 'Influence of turbulence closure on free-surface flow modelling with SPH', *XXXII IAHR Biennial Congress*. Venice (Italy).

Violeau, D. and Issa, R. (2007b) 'Numerical modelling of complex turbulent free-surface flows with the SPH method: an overview', *International Journal for Numerical Methods in Fluids*, 53 (2), pp.277-304.

Volz, C., Rousselot, P. and Vetsch, D. (2011) 'Numerical Modelling of Earth Embankment Breaching Processes Due to Overtopping Flow on Unstructured Meshes', *Journal of Hydraulic Engineering-Asce*, (submitted).

Wilcock, P. R. and Crowe, J. C. (2003) 'Surface-based transport model for mixed-size sediment', *Journal of Hydraulic Engineering-Asce*, 129 (2), pp.120-128.

Wolf-Gladrow, D. (2000) *Lattice-gas cellular automata and lattice boltzmann models: an introduction*. Springer, Berlin.

Wong, M. G. and Parker, G. (2006) 'Reanalysis and correction of bed-load relation of Meyer-Peter and Muller using their own database', *Journal of Hydraulic Engineering-Asce*, 132 (11), pp.1159-1168.

Wu, F. C. and Chou, Y. J. (2003) 'Rolling and lifting probabilities for sediment entrainment', *Journal of Hydraulic Engineering-Asce*, 129 (2), pp.110-119.

Wu, F. C. and Lin, Y. C. (2002) 'Pickup probability of sediment under log-normal velocity distribution', *Journal of Hydraulic Engineering-Asce*, 128 (4), pp.438-442.

Wu, F. C. and Yang, K. H. (2004) 'Entrainment probabilities of mixed-size sediment incorporating near-bed coherent flow structures', *Journal of Hydraulic Engineering-Asce*, 130 (12), pp.1187-1197.

Wu, W. M. (2008) *Computational river dynamics*. Taylor & Francis, London.

Wu, W. M., Rodi, W. and Wenka, T. (2000a) '3D numerical modeling of flow and sediment transport in open channels', *Journal of Hydraulic Engineering-Asce*, 126 (1), pp.4-15.

Wu, W. M., Wang, S. S. Y. and Jia, Y. F. (2000b) 'Nonuniform sediment transport in alluvial rivers', *Journal of Hydraulic Research*, 38 (6), pp.427-434.

Xiong, Q., Deng, L., Wang, W. and Ge, W. (2011) 'SPH method for two-fluid modeling of particle-fluid fluidization', *Chemical Engineering Science*, 66 (9), pp.1859-1865.

Yalin, M. S. (1977) *Mechanics of sediment transport*. 2nd edn. Pergamon Press, Toronto.

Yalin, M. S. and da Silva, A. M. F. (2001) *Fluvial processes*. IAHR International Association of Hydraulic Engineering and Research.

Yalin, M. S. and Karahan, E. (1979) 'Inception of sediment transport', *Journal of the Hydraulics Division-Asce*, 105 (11), pp.1433-1443.

Yang, S. Q. (2005) 'Prediction of total bed material discharge', *Journal of Hydraulic Research*, 43 (1), pp.12-22.

Young, D. C. (2002) 'Molecular Mechanics', in *Computational Chemistry*. John Wiley & Sons, pp.49-59.

Yu, Z. and Fan, L.-S. (2010) 'Lattice Boltzmann method for simulating particle-fluid interactions', *Particuology*, 8 (6), pp.539-543.

Zanke, U. C. E. (1982) *Grundlagen der Sedimentbewegung*. Springer, Berlin.

Zedler, E. A. and Street, R. L. (2001) 'Large-eddy simulation of sediment transport: Currents over ripples', *Journal of Hydraulic Engineering-Asce*, 127 (6), pp.444-452.

Zhou, Y. C., Wright, B. D., Yang, R. Y., Xu, B. H. and Yu, A. B. (1999) 'Rolling friction in the dynamic simulation of sandpile formation', *Physica A: Statistical Mechanics and its Applications*, 269 (2-4), pp.536-553.

Zohdi, T. I. (2005) 'Charge-induced clustering in multifield particulate flows', *International Journal for Numerical Methods in Engineering*, 62 (7), pp.870-898.

List of Symbols

Latin Capitals

A	[$m^{\sigma-1}$]	surface area
A_I	[.]	integral interpolant of any quantity or function A_r
A_o	[m]	amplitude of harmonic oscillator
A_r	[.]	any quantity or function
$A_\perp$	[$m^{\sigma-1}$]	projected area for drag force
$A_\parallel$	[$m^{\sigma-1}$]	projected area for lift force
Ar	[-]	Archimedes number
B	[N/$m^{\sigma-1}$]	coefficient in equation of state
B_*	[-]	constant in Einstein formula
C'	[N]	cohesion
C_D	[-]	drag coefficient
C_L	[-]	lift coefficient
C_w	[-]	weir coefficient
D_*	[-]	dimensionless grain diameter
D_w	[m]	influence distance from wall of MLJ potential
$\hat{D}$	[kg/(sm^3)]	discretised form of continuity equation
E	[J/kg]	total energy per unit mass
E_{tot}	[J]	total energy
E_k	[N/m^2]	Young's modulus
$\hat{F}$	[m^2/s^2]	discretised form of momentum equation
$\vec{F}_a$	[N]	applied forces
$\vec{F}_b$	[N]	buoyancy force
$\vec{F}_c$	[N]	contact force
$\vec{F}_d$	[N]	damping force
$\vec{F}_{dr}$	[N]	drag force
$\vec{F}_D$	[N]	total dissipative force
$\vec{F}_{D,t}$	[N]	tangential dissipative force
$\vec{F}_g$	[N]	weight of body
$\vec{F}_{g*}$	[N]	submerged weight of body
$\vec{F}_{gh}$	[N]	sum of hydrostatic pressure force and weight of body
$\vec{F}_h$	[N]	hydrostatic pressure force
$\vec{F}_l$	[N]	lift force
$\vec{F}_n$	[N]	normal force
$\vec{F}_{Rk}$	[N]	kinetic friction force
$\vec{F}_{Rs}$	[N]	static friction force

List of Symbols

Symbol	Unit	Description
$\vec{F}_{Rv}$	[N]	viscous friction force
$\vec{F}_s$	[N]	spring force
$\vec{F}_{s,t}$	[N]	force of tangential spring
$\vec{F}_t$	[N]	tangential force
H_0	[m]	initial water depth
$\mathbf{I}$	[kg m^2]	tensor of moment of inertia
$\mathbf{I}_1$	[-]	unit tensor
K	[N/m]	generalised stiffness constant, material parameter of Hertz law
$\vec{L}$	[N m s]	angular momentum
L_s	[m]	non-equilibrium adaption length
$\vec{M}_a$	[N m]	applied torques
$\vec{M}_c$	[N m]	contact torque
$\vec{M}_e$	[N m]	external torque
$\vec{M}_f$	[N m]	friction torque
Ma	[-]	Mach number
$\mathbf{Q}$	[-]	orthogonal quaternion matrix
Q_b	[m^3/s]	non-equilibrium bed load
Q_e	[m^3/s]	equilibrium transport rate
Q_Γ	[1/m$^{\sigma\text{-}1}$]	boundary source vector
Q_V	[1/m$^\sigma$]	volume source
R	[m]	influence distance of LJ potential
R_{ab}	[m^2/s^2]	influence distance of LJ potential
R_r	[-]	coefficient of roundness
Re	[-]	Reynolds number
Re*	[-]	grain or bed particle Reynolds number
Re$^*_c$	[-]	critical particle Reynolds number
S_b	[-]	channel or bed slope
S_e	[-]	slope of energy grade line
$S_{f,g}$	[m^2/s]	active stratum source term
T	[K]	temperature
T_o	[s]	period of harmonic oscillator
U	[J]	potential energy
U_s	[J]	potential energy of displaced spring
U_h	[J]	potential energy of Hertz force law
V	[m$^\sigma$]	volume of body
V_s	[m$^\sigma$]	volume of sphere
W	[1/m$^\sigma$]	kernel function
W_{ab}	[1/m$^\sigma$]	average of kernel functions
W_f	[J/kg]	work of external forces per unit mass

List of Symbols

Latin Minuscules

a	[m]	measure of displacement
a_t	[m/s^2]	tangential acceleration
a_{max}	[m/s^2]	maximum particle acceleration
$\vec{a}_i$	[m/s^2]	acceleration of particle i
c	[N/m]	stiffness of spring
$\overline{c}_{ab}$	[m/s]	averaged sound velocity of SPH particles a and b
c_f	[-]	Chézy coefficient
c_R	[-]	coefficient of logarithmic law for small relative flow depth
c_s	[m/s]	sound velocity
c_t	[N/m]	stiffness of tangential spring
c_v	[J/(kg K)]	specific heat capacity
d	[Ns/m]	viscous damping coefficient
d_t	[Ns/m]	tangential damping coefficient
d_s	[m]	grain diameter
e	[J/kg]	internal energy per unit mass
e_r	[-]	restitution factor
$\vec{e}$	[-]	unit vector
$\vec{e}_{ij}$	[-]	interaction unit vector
$\vec{e}_{sd}$	[-]	unit vector of the spring-damper system
$\vec{e}_t$	[-]	tangential unit vector
f_{ab}	[-]	kernel dependant scaling function
$\vec{f}_e$	[N/m$^\sigma$]	external volume forces`
f_0	[1/s]	ordinary frequency
g	[m/s^2]	gravitational acceleration
$\vec{g}$	[m/s^2]	gravitational acceleration vector
h	[m]	smoothing length of SPH kernel
h_a	[m]	smoothing length of SPH particle a
h_f	[m]	flow depth
h_m	[m]	height of active layer
h_w	[m]	weir height
k	[N]	stiffness of LJ potential
k_b	[m$^{1/3}$/s]	Manning-Strickler coefficient including bed forms
k_r	[m$^{1/3}$/s]	Manning-Strickler coefficient based on grain diameter
m	[kg]	mass
m_s	[kg]	mass of sphere
m_w	[kg]	mass of boundary particle
n_g	[-]	number of grain size classes
p	[N/m$^{\sigma-1}$]	pressure
p_a	[N/m$^{\sigma-1}$]	pressure of SPH particle a
p_e	[-]	pickup probability
q_b	[m^2/s]	volume bed load transport rate per unit channel width
$q_{b,g}$	[m^2/s]	bed load rate per unit channel width of grain size class g
q_H	[1/m$^\sigma$]	sources other than conduction

List of Symbols

$\mathbf{q}_i$	[-]	rotation unit quaternion of particle i
q_v	[N/m^2]	contribution to artificial viscosity term
q_w	[m^2/s]	water discharge per unit channel width with sidewall correction
q'_w	[m^2/s]	water discharge per unit channel width without correction
q_{we}	[m$^\sigma$/s]	specific weir discharge
r	[m]	distance from origin
r_{ab}	[m]	distance between SPH particles a and b
$\vec{r}_a$	[m]	position vector of SPH particle a
r_i	[m]	radius of particle i
$\vec{r}_i$	[m]	position vector of particle i
r_{ij}	[m]	distance between particles i and j
$r_{\min}$	[m]	radius of smallest particle of simulation
r_s	[m]	sphere radius
r_0	[m]	point where the LJ force changes from repulsion to attraction
s	[-]	sediment specific density, $s = \rho_s/\rho_f$
t_c	[s]	collision time
u_*	[m/s]	shear stress velocity
u_{*c}	[m/s]	critical shear stress velocity
$\vec{u}$	[m/s]	velocity vector
$\vec{u}_a$	[m/s]	velocity vector of SPH particle a
$\vec{u}_f$	[m/s]	free stream velocity
$\overline{u}_m$	[m/s]	depth-averaged flow velocity
u_{ref}	[m/s]	reference velocity
v_c	[m/s]	contact or impact velocity
v'_c	[m/s]	velocity after collision
v_{cs}	[m/s]	creep speed
v_{end}	[m/s]	terminal velocity
$\vec{v}_i$	[m/s^2]	velocity of particle i
v_t	[m/s]	tangential velocity
v_0	[m/s]	initial velocity
w_s	[m/s]	terminal settling velocity
$\hat{w}_s$	[m/s]	reduced terminal settling velocity
$\overline{w}_s$	[m/s]	observed terminal settling velocity
y_R	[m]	height of roughness sublayer
z_b	[m]	bed level
z_0	[m]	datum

List of Symbols

Greek Symbols

α	[-]	artificial viscosity coefficient for volume-viscous pressure
α_g	[-]	coefficient for kernel gradient force law
α_s	[-]	safety factor for time step conditions
α_σ	[$1/m^\sigma$]	coefficient for Gaussian kernel function
β	[-]	artificial viscosity coefficient for Von Neumann-Richtmyer pressure
β_k	[-]	coefficient for kernel gradient force law
β_g	[-]	volume fraction of grain size class g
Γ	[$m^{\sigma-1}$]	domain boundary
Γ_g	[m/s^2]	gradient force function
Δs	[m]	initial particle spacing
Δt	[s]	size of time step
γ	[°]	inclination of channel bed
γ_a	[-]	adiabatic index
γ_d	[-]	attenuation factor or damping ratio
γ_p	[-]	exponent in equation of state
γ_{pc}	[-]	damping coefficient of PC-leapfrog scheme
γ_w	[-]	reduction coefficient to account for wall interference
δ	[m]	penetration depth or displacement of spring
δ_t	[m]	displacement of tangential spring
δ_w	[m]	distance to boundary or wall
δ_{weq}	[m]	equilibrium distance to boundary or wall
δ_{w0}	[m]	distance from wall of zero force point of MLJ potential
δ_0	[m]	offset from equilibrium position
ε_p	[Nm]	stiffness of Lennard-Jones potential
ε_s	[-]	artificial stress coefficient
ε_X	[-]	XSPH coefficient
η	[-]	friction slope
η_v	[m]	term for artificial viscosity to prevent singularities
η_0	[-]	standard deviation in Einstein formula
θ	[°]	phase angle
θ_c	[-]	dimensionless critical shear stress, Shields parameter
κ	[-]	von Kármán constant
λ_w	[-]	particle to tank width ratio
μ	[Ns/m^2]	dynamic viscosity
μ_{ab}	[m/s]	term for artificial viscosity
μ_k	[-]	kinetic friction coefficient
μ_m	[-]	reduced or effective mass
μ_s	[-]	static friction coefficient
μ_v	[-]	viscous friction coefficient
μ_w	[-]	weir discharge coefficient
ν	[m^2/s]	kinematic viscosity
ν_k	[-]	Poisson's ratio
Ξ	[-]	Yalin parameter

List of Symbols

Π_{ab}	[N/m^2]	artificial viscosity term
ρ_a	[kg/m^3]	density of SPH particle a
$\overline{\rho}_{ab}$	[kg/m^3]	averaged density of SPH particles a and b
ρ_0	[kg/m^3]	initial or reference density of the fluid
ρ_f	[kg/m^3]	density of the fluid
ρ_s	[kg/m^3]	density of the sediment or granular material
ρ_w	[kg/m^3]	density of boundary particle
$\triangle\rho$	[kg/m^3]	density difference between sediment and fluid
$\boldsymbol{\sigma}$	[N/m^2]	internal stress tensor
σ	[-]	dimension
σ_k	[-]	material parameter of Hertz force law
σ_p	[-]	parameter of Lennard-Jones potential
$\boldsymbol{\tau}$	[N/m^2]	viscous stress tensor
τ_{ij}	[N/m^2]	shear stress
τ_b	[N/m^2]	bed shear stress
τ_b^*	[N/m^2]	dimensionless bed shear stress
τ_c	[N/m^2]	critical bed shear stress
τ_c^*	[-]	dimensionless critical shear stress, Shields parameter
ϕ	[deg]	angle of repose
ϕ_{cv}	[deg]	constant volume critical state friction angle
$\vec{\Phi}$	[1/m$^\sigma$]	flux vector
$\vec{\Phi}_C$	[1/m$^\sigma$]	convective flux vector
$\vec{\Phi}_D$	[1/m$^\sigma$]	diffusive flux vector
$\boldsymbol{\Phi}$	[1/m$^\sigma$]	flux tensor
$\boldsymbol{\Phi}_C$	[1/m$^\sigma$]	convective flux tensor
$\boldsymbol{\Phi}_D$	[1/m$^\sigma$]	diffusive flux tensor
χ	[-]	hysteresis factor
ψ	[deg]	dilation angle
ψ_R	[-]	ratio between tangential force and maximum viscous friction force
ψ_{eq}	[-]	force potential parameter
Ψ	[1/m$^\sigma$]	quantity per unit volume
Ψ_*	[m^2/N]	flow intensity
$\vec{\Psi}$	[1/m$^\sigma$]	vector quantity per unit volume
Ω	[m$^\sigma$]	domain
$\boldsymbol{\Omega}_i$	[-]	pure quaternion of angular velocity
ω_0	[1/s]	angular frequency
ω	[rad/s]	angular velocity
$\vec{\omega}$	[rad/s]	angular velocity vector

General Mathematical Expressions and Symbols

t	[s]	time
x	[m]	Cartesian coordinate, main flow direction
y	[m]	Cartesian coordinate, horizontally perpendicular to x
z	[m]	Cartesian coordinate, vertical direction in general
$\hat{x}$	[N]	coordinate parallel to channel bottom
$\hat{z}$	[N]	coordinate perpendicular to channel bottom
ζ	[m]	coordinate for harmonic oscillator
$\delta(r)$		Dirac delta function
δ_{ij}		Kronecker delta, $\delta_{ij} = 1$ for $i = j$, otherwise $\delta_{ij} = 0$
$|\vec{x}|$		Euclidean norm of vector $\vec{x}$
$\{P\}$	[-]	numerical quantity of P
$\vec{\nabla}_a W_{ab}$		kernel gradient with respect to SPH particle a
$\vec{\nabla}$		nabla operator: $\vec{\nabla} = (\partial/\partial x_1, \partial/\partial x_2, \partial/\partial x_3)$
Δ		Laplacian: $\Delta = \vec{\nabla}^2 = \vec{\nabla} \cdot \vec{\nabla} = \sum_{j=1}^{3} \partial^2/\partial x_j^2$

List of Acronyms

Bi-CGSTAB	BiConjugate Gradient Stabilised Method
DEM	Discrete/Distinct Element Method
DES	Detached Eddy Simulation
DNS	Direct Numerical Simulation
FAVOR	Fractional-Area-Volume Obstacle Representation
FDM	Finite Difference Method
FEM	Finite Element Method
FSI	Fluid-Structure Interaction
FVM	Finite Volume Method
GPU	Graphics Processing Unit
HRFM	High Resolution Force Model
ISPH	truly Incompressible Smoothed Particle Hydrodynamics
LBM	Lattice Boltzmann Method
LDV	Laser Doppler Velocimetry
LES	Large Eddy Simulation
LJ	Lennard-Jones
LRFM	Low Resolution Force Model
MAC	Marker-And-Cell
MLJ	Modified Lennard-Jones
PIC	Particle-In-Cell
Pasimodo	PArticle SImulation and MOlecular Dynamics in an Object oriented fashion
RANS	Reynolds-Averaged Navier-Stokes
rSPH	remeshed Smoothed Particle Hydrodynamics
SPH	Smoothed Particle Hydrodynamics
VOF	Volume Of Fluid
WCSPH	Weakly Compressible Smoothed Particle Hydrodynamics

Appendix

A.1 Material Properties

A.1.1 Stiffness and Young's Modulus

For an isotopic elastic material, Hooke's law holds,

$$\sigma = E\varepsilon \tag{8.1}$$

where the Young's modulus is defined as

$$E = \frac{\sigma}{\varepsilon} = \frac{F_A / A_0}{\Delta l / l_0}, \tag{8.2}$$

with σ the tensile normal stress, ε the tensile normal strain and the force $F_A = \vec{F}_A \cdot \vec{n}$ acting in the normal direction on the cross-sectional area A_0 leading to an increase/decrease of the original length l_0 by Δl. Appropriate values for Young's modulus of real materials can be found in the literature, e.g. from Ashby and Jones (2005).

Equation (8.2) can be rearranged for a linear elastic spring with force $F_s = F_A$, stiffness $c = E\,A_0/l_0$ and displacement $\delta = \Delta l$:

$$F_s = c\delta. \tag{8.3}$$

In practice, the constant c is usually determined by the physical properties of the spring.

A.1.2 Poisson Ratio

For a linear elastic, and density preserving material the Poisson ratio is $\nu = 0.5$, i.e. deformation does not cause a change in volume and the material is termed "elastic incompressible". For real materials the Poisson ratio is $\nu < 0.5$, e.g. the mean value for metal is approximately 0.3 and for granite approximately 0.25. Details on the determination of the Poisson ratio for rocks can be found in Burshtein (1968). According to Mitchell and Soga (2005) the Poisson ration for geomaterials varies in the range of 0 – 0.35.

A.1.3 Viscosity

According to Newton's law of viscosity, the shear stress acting between two parallel fluid layers is defined as

$$\tau = \mu \frac{du}{dy} , \qquad (8.4)$$

where u is the velocity component perpendicular to y and μ is the dynamic viscosity of the fluid, a material property,

$$\mu = \frac{\text{Force}}{\text{Area}} \bigg/ \frac{\text{Velocity}}{\text{Distance}} = \frac{[\text{N}][\text{s}]}{[\text{m}^2]} . \qquad (8.5)$$

The ratio of the dynamic viscosity to the mass density defines the kinematic viscosity of the fluid:

$$\nu = \frac{\mu}{\rho} \qquad (8.6)$$

A.1.4 Friction Coefficients of Granular Materials

The coefficient of static friction, μ_s of a granular material can be approximated by its angle of repose ϕ (also termed angle of internal friction),

$$\mu_s = \tan(\phi) . \qquad (8.7)$$

According to Santamarina and Cho (2004), the angle of repose depends on the roundness of the particles, and the following linear fit for the constant volume critical state friction angle ϕ_{cv} is proposed

$$\phi_{cv} = 42 - 17 R_r , \qquad (8.8)$$

where R_r is the coefficient of roundness defined in Mitchell and Soga (2005), with $R_r = 1$ for spherical particles. A similar result was obtained by Sukumaran and Ashmawy (2001). They present friction angles from drained triaxial tests for different particle shapes. Furthermore, for a granular material of dense packing the dilation can be an important component of the shear strength (see e.g. Santamarina and Cho (2004)). Thus, the peak friction angle can be expressed by

$$\phi_{max} = \phi_{cv} + 0.8 \psi , \qquad (8.9)$$

where ψ is the dilation angle. E.g. for locked sands values of the dilation angle about 30° have been measured. The effect of dilation on the shear strength is also known as cohesion and the corresponding shear or friction force reads

$$\vec{F}_s = \vec{F}_N \tan(\phi_{cv}) + C', \qquad (8.10)$$

where C' is the cohesion that is usually set to zero for granular materials with large grain sizes (see Potapov and Campbell (1996)). This seems to be reasonable since the granular material is modelled by discrete particles and hence dilation can explicitly take place. Therefore only the part for constant volume friction is of relevance, i.e. the static friction coefficient of the modelled granular material is $\mu_s = \tan(\phi_{cv})$. Furthermore it is assumed, that the coefficient of kinematic friction is equal to that of static friction, i.e. $\mu_k = \mu_s$. For a given roundness $R_r \approx 0.7$ of the grains, equations (8.8) and (8.7) result in $\phi = 30°$ and $\mu_s = 0.58$. Schellart (2000) observed that the physical handling technique to deposit the sediment piles has a major influence on the angle of repose. According to his results, the former values are rather at the lower boundary. For very well rounded grains with very high sphericity values of $\phi = 41°$ and $\mu_s = 0.87$ were also observed.

Due to the inconsistency reported above, intermediate values are suggested for the present work,

$$\phi = 35°, \quad \mu_s = \mu_k = 0.7 \ . \qquad (8.11)$$

A.1.5 Summary

The subsequently listed values are used for this work and are approximate values. They may vary depending on the composition of the material or environmental influences such as temperature.

Bed material: gravel consisting of granite

property	symbol	value	units
density	ρ_s	2800	kg/m³
Young's modulus	E_k	60 x10⁹	N/m²
Poisson ratio	ν_k	0.25	-
static friction coefficient	μ_s	0.7	-
kinetic friction coefficient	μ_k	0.7	-
rolling resistance coefficient	μ_r	0.001	-

Fluid: water at 20°celsius

property	symbol	value	units
density	ρ_f	998.21	kg/m³
dynamic viscosity	μ	1.0 x10⁻³	Ns/m²
kinematic viscosity	ν	1.01 x10⁻⁶	m²/s
sound velocity	c_s	1484	m/s

A.2 General Definitions for DEM

The following definitions are used for the description of the discrete element method in chapter 4.3. Consider two particles i and j with position vectors $\vec{r}_i$ and $\vec{r}_j$. Note that in the used model also rigid walls are made of particles, i.e. triangles.

The distance vector between two particles always depends on the particle considered at a time,

$$\vec{r}_{ij} := \begin{cases} (\vec{r}_b - \vec{r}_a) & \text{for particle } a, \\ (\vec{r}_a - \vec{r}_b) & \text{for particle } b, \end{cases} \qquad (8.12)$$

and the distance between the two particles is given by the Euclidean norm

$$r_{ij} := \left| \vec{r}_{ij} \right|. \qquad (8.13)$$

For spheres, where the distance vector is normal to the contact surface, the interaction unit vector can be defined as

$$\vec{e}_{ij} := \frac{\vec{r}_{ij}}{r_{ij}}. \qquad (8.14)$$

Note that for arbitrarily shaped bodies the interaction unit vector has to be derived from the contact normal and not from the connection between the centres.

The velocities of the particles are defined as $\vec{v}_i$ and $\vec{v}_j$. In compliance with equation (8.13) the norm of the velocity reads

$$v_i := \left| \vec{v}_i \right|. \qquad (8.15)$$

For a collision $\vec{v}_i \neq \vec{v}_j$ holds. The velocity components in normal and tangential direction of the collision are obtained by projection onto the interaction unit vector. Thus, the interaction velocity is defined as

$$\vec{v}_{ij} = \frac{\vec{v}_i \cdot \vec{r}_{ij}}{r_{ij}} \vec{e}_{ij} = \left(\vec{v}_i \cdot \vec{e}_{ij} \right) \vec{e}_{ij}. \qquad (8.16)$$

Similar relations hold for particle j. The relative velocity of the two particles is

$$\vec{v}_r = \vec{v}_i - \vec{v}_j \qquad (8.17)$$

and the relative tangential velocity is obtained by

Appendix

$$\vec{v}_t = \vec{v}_r - \left(\vec{v}_r \cdot \vec{e}_{ij}\right)\vec{e}_{ij} \,, \tag{8.18}$$

leading to the tangential unit vector

$$\vec{e}_t = \frac{\vec{v}_t}{v_t} \,, \tag{8.19}$$

where $v_t := \left|\vec{v}_t\right|$.

A.3 MLJ Potential for Sphere to Sphere Interaction

For the interaction of a fluid particle with a sphere of radius r_s the following transformation is defined

$$\text{fluid to sphere}: \quad \delta_w := r_{ij} - r_s,\ \delta_{w0} := r_0 - r_s,\ D_w := R - r_s\ . \tag{8.20}$$

By insertion of (8.20) into (4.116) one obtains

$$\vec{F}_n(\delta_w) = \begin{cases} k\dfrac{(\delta_w - D_w)^2(\delta_w - \delta_{w0})(\delta_w - 2D_w + \delta_{w0})}{(D_w + r_s)^2(2D_w - \delta_{w0} + r_s)(\delta_{w0} + r_s)}\vec{e}_{ij}, & \text{if } \delta_w < D_w, \\ 0 & \text{otherwise}, \end{cases} \tag{8.21}$$

and the corresponding potential reads

$$U(\delta_w) = \frac{k}{(D_w + r_s)^2(2D_w - \delta_{w0} + r_s)(\delta_{w0} + r_s)}\left[\frac{(\delta_w + r_s)^5}{5} - (\delta_w + r_s)^4(D_w + r_s)\right.$$
$$+(\delta_w + r_s)(D_w + r_s)^2(2D_w - \delta_{w0} + r_s)(\delta_{w0} + r_s)$$
$$-(\delta_w + r_s)^2(D_w + r_s)((D_w + r_s)^2 + 2(D_w + r_s)(\delta_{w0} + r_s) - (\delta_{w0} + r_s)^2) \tag{8.22}$$
$$\left.\frac{(\delta_w + r_s)^3}{3}(5(D_w + r_s)^2 + 2(D_w + r_s)(\delta_{w0} + r_s) - (\delta_{w0} + r_s)^2)\right],$$
$$\text{if } \delta_w < D_w,\ \text{otherwise } U(\delta_w) = 0\ .$$

Considering only repulsive forces, i.e. $D_w = \delta_{w0}$, equation (8.22) becomes

$$\vec{F}_n(\delta_w) = \frac{k}{(\delta_{w0} + r_s)^4}(\delta_{w0} - \delta_w)^4\vec{e}_{ij}, \quad \delta_w \le \delta_{w0}\ . \tag{8.23}$$

The corresponding potential reads

$$U(\delta_w) = \frac{k}{(\delta_{w0} + r_s)^4}\Big((\delta_w + r_s)^5\big/5 - (\delta_w + r_s)^4(\delta_{w0} + r_s) + 2(\delta_w + r_s)^3(\delta_{w0} + r_s)^2$$
$$-2(\delta_w + r_s)^2(\delta_{w0} + r_s)^3 + (\delta_w + r_s)(\delta_{w0} + r_s)^4\Big),\ \delta_w \le \delta_{w0}\ . \tag{8.24}$$

Appendix

The maximum of the repulsive-only potential is at distance δ_{w0} from the wall and has the value

$$U(\delta_{w0}) = \frac{1}{5} k \left(\delta_{w0} + r_s \right) . \tag{8.25}$$

To balance a given external static force of amount F, e.g. the weight mg of a fluid body with mass m, the equilibrium distance to the wall can be determined by

$$\delta_{weq} = \delta_{w0} - \left(\delta_{w0} + r_s \right) \left(\frac{F}{k} \right)^{1/4} \tag{8.26}$$

and the corresponding potential is

$$U(\delta_{weq}) = \frac{\delta_{w0} + r_s}{5} \left[k - F \left(\frac{F}{k} \right)^{1/4} \right] . \tag{8.27}$$

If a certain equilibrium distance to the wall is preferred, the appropriate stiffness could be obtained by rearranging (8.26),

$$k = F \left(\frac{\delta_{w0} + r_s}{\delta_{w0} - \delta_{weq}} \right)^4 . \tag{8.28}$$

The minimal stiffness $k_{\min}$ to prevent penetration of an approaching body with mass m and velocity u_w is

$$k_{\min} = \frac{5m}{\delta_{w0} + r_s} \left(g\delta_{w0} + \frac{u_w^2}{2} \right) . \tag{8.29}$$

For a potential with $\delta_{weq} = \delta_{w0}/2$, equation (8.28) can be written as

$$k = 16F \left(\frac{d_{w0} + r_s}{d_{w0}} \right)^4 ; \quad F = \frac{k}{16} \left(\frac{d_{w0}}{d_{w0} + r_s} \right)^4 . \tag{8.30}$$

A.4 Different Forms of Lennard-Jones Potentials

A.4.1 Form Depending on Equilibrium Distance

Inserting equation $\sigma = 2^{-1/6} r_0$ into (4.75) leads to

$$\vec{F}_n(r_{ij}) = 12\varepsilon \frac{1}{r_{ij}} \left[\left(\frac{r_0}{r_{ij}}\right)^6 - \left(\frac{r_0}{r_{ij}}\right)^{12} \right] \vec{e}_{ij} \, . \tag{8.31}$$

A.4.2 Form According to Monaghan

Monaghan (1994) uses as impermeable boundary condition a Lennard-Jones potential of the form

$$\vec{F}_n(r_{ij}) = \frac{k}{r_{ij}} \left[\left(\frac{r_0}{r_{ij}}\right)^{p_1} - \left(\frac{r_0}{r_{ij}}\right)^{p_2} \right] \vec{e}_{ij} \, , \tag{8.32}$$

with corresponding potential

$$U(r_{ij}) = k \left(\frac{1}{p_2} \left(\frac{r_0}{r_{ij}}\right)^{p_2} - \frac{1}{p_1} \left(\frac{r_0}{r_{ij}}\right)^{p_1} \right) . \tag{8.33}$$

He sets $p_1 = 4$ and $p_2 = 2$ for most of his simulations, but found similar results for $p_1 = 12$ and $p_2 = 6$.

A.4.3 Form According to Müller

The Lennard-Jones potential according to the force law presented by Muller *et al.* (2004) can be obtained by integration of equation (4.116); one obtains

$$U(r_{ij}) = \frac{k}{R^2(2R - r_0)r_0} \left[\frac{r_{ij}^5}{5} - r_{ij}^4 R + r_{ij} R^2 (2R - r_0) r_0 \right.$$
$$\left. - r_{ij}^2 R(R^2 + 2Rr_0 - r_0^2) \right.$$
$$\left. + \frac{r_{ij}^3}{3}(5R^2 + 2Rr_0 - r_0^2) \right], \tag{8.34}$$
$$\text{if } r_{ij} < R, \text{ otherwise } U(r_{ij}) = 0 \, .$$

Considering only repulsive forces, i.e. $R = r_0$, equation (4.116) becomes

Appendix

$$\vec{F}_n(r_{ij}) = \frac{k}{r_0^4}(r_0 - r_{ij})^4 \vec{e}_{ij} \ . \tag{8.35}$$

The corresponding potential reads

$$U(r_{ij}) = \frac{k}{r_0^4}\left(\frac{r_{ij}^5}{5} - r_{ij}^4 r_0 + 2r_{ij}^3 r_0^2 - 2r_{ij}^2 r_0^3 + r_{ij} r_0^4\right) \ . \tag{8.36}$$

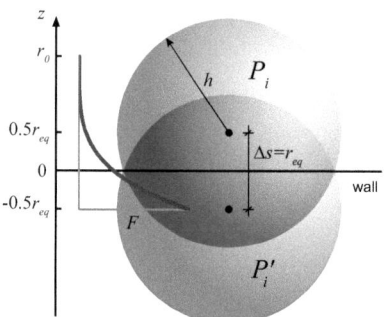

Fig. A-1: Lennard-Jones potential as boundary force law with the particle resting at equilibrium position.

To balance a given external force of amount F, e.g. the weight mg of a body with mass m, the equilibrium position can be determined by

$$r_{eq} = r_0\left(1 - \left(\frac{F}{k}\right)^{1/4}\right) \ . \tag{8.37}$$

Furthermore, rearranging (8.37) yields

$$k = F\left(\frac{r_0}{r_0 - r_{eq}}\right)^4 \ . \tag{8.38}$$

Defining a potential with $r_{eq} = r_0/2$ leads to

$$k = 16F, \ U(r_0) = \frac{1}{5}kr_0, \ U(r_{eq}) = \frac{31}{160}kr_0 \ . \tag{8.39}$$

The recommended SPH parameters are (see Fig. A-1):

$$\Delta s = r_{eq}, \quad h = 1.5\Delta s \ . \tag{8.40}$$

A.4.4 Approximations

The polynomial (4.121) describing the potential $U(d_w)$ can be approximated by

$$U(d_w) \approx \tanh\left(4.1\frac{d_w}{d_{w0}}\right) U(d_{w0}), \quad d_w \leq d_{w0} \ , \tag{8.41}$$

where $U(d_{w0})$ is given by equation (4.122).

A.5 Vector Projection

Consider a vector $\vec{r} = (0\ 0\ z)_{xyz}$ given in a Cartesian coordinate system with axes x, y, z and a similar coordinate system with axes $\hat{x}, \hat{y}, \hat{z}$ that is clockwise rotated around the y-axis by angle γ, i.e. γ is negative. The projection of $\vec{r}$ onto the axis of $\hat{x}, \hat{y}, \hat{z}$ reads

$$\vec{r}_{\hat{x}} = \begin{vmatrix} 0 & 0 & \sin\gamma\cos\gamma \\ 0 & 0 & 0 \\ 0 & 0 & \sin^2\gamma \end{vmatrix} \cdot \vec{r} = z \begin{pmatrix} \sin\gamma\cos\gamma \\ 0 \\ \sin^2\gamma \end{pmatrix},$$

$$\vec{r}_{\hat{z}} = \begin{vmatrix} 0 & 0 & -\sin\gamma\cos\gamma \\ 0 & 0 & 0 \\ 0 & 0 & \cos^2\gamma \end{vmatrix} \cdot \vec{r} = z \begin{pmatrix} -\sin\gamma\cos\gamma \\ 0 \\ \cos^2\gamma \end{pmatrix}. \tag{8.42}$$

The corresponding unit vectors are

$$\vec{e}_{\hat{x}} = \begin{pmatrix} \cos\gamma \\ 0 \\ \sin\gamma \end{pmatrix}; \quad \vec{e}_{\hat{z}} = \begin{pmatrix} -\sin\gamma \\ 0 \\ \cos\gamma \end{pmatrix}. \tag{8.43}$$

i want morebooks!

Buy your books fast and straightforward online - at one of world's fastest growing online book stores! Environmentally sound due to Print-on-Demand technologies.

Buy your books online at
www.get-morebooks.com

Kaufen Sie Ihre Bücher schnell und unkompliziert online – auf einer der am schnellsten wachsenden Buchhandelsplattformen weltweit! Dank Print-On-Demand umwelt- und ressourcenschonend produziert.

Bücher schneller online kaufen
www.morebooks.de

 VDM Verlagsservicegesellschaft mbH
Heinrich-Böcking-Str. 6-8   Telefon: +49 681 3720 174   info@vdm-vsg.de
D - 66121 Saarbrücken   Telefax: +49 681 3720 1749   www.vdm-vsg.de

Printed by Books on Demand GmbH, Norderstedt / Germany